Conversion Factors

Mass

1 g = 10^{-3} kg
1 kg = 10^{3} g
1 u = 1.66×10^{-24} g = 1.66×10^{-27} kg
1 metric ton = 1000 kg

Length

1 nm = 10^{-9} m
1 cm = 10^{-2} m = 0.394 in.
1 m = 10^{-3} km = 3.28 ft = 39.4 in.
1 km = 10^{3} m = 0.621 mi
1 in. = 2.54 cm = 2.54×10^{-2} m
1 ft = 0.305 m = 30.5 cm
1 mi = 5280 ft = 1609 m = 1.609 km

Area

1 cm^2 = 10^{-4} m^2 = 0.1550 in^2 = 1.08×10^{-3} ft^2
1 m^2 = 10^{4} cm^2 = 10.76 ft^2 = 1550 in^2
1 in^2 = 6.94×10^{-3} ft^2 = 6.45 cm^2 = 6.45×10^{-4} m^2
1 ft^2 = 144 in^2 = 9.29×10^{-2} m^2 = 929 cm^2

Volume

1 cm^3 = 10^{-6} m^3 = 3.35×10^{-5} ft^3 = 6.10×10^{-2} in^3
1 m^3 = 10^{6} cm^3 = 10^{3} L = 35.3 ft^3 = 6.10×10^{4} in^3 = 264 gal
1 liter = 10^{3} cm^3 = 10^{-3} m^3 = 1.056 qt = 0.264 gal
1 in^3 = 5.79×10^{-4} ft^3 = 16.4 cm^3 = 1.64×10^{-5} m^3
1 ft^3 = 1728 in^3 = 7.48 gal = 0.0283 m^3 = 28.3 L
1 qt = 2 pt = 946 cm^3 = 0.946 L
1 gal = 4 qt = 231 in^3 = 0.134 ft^3 = 3.785 L

Time

1 h = 60 min = 3600 s
1 day = 24 h = 1440 min = 8.64×10^{4} s
1 y = 365 days = 8.76×10^{3} h = 5.26×10^{5} min = 3.16×10^{7} s

Angle

1 rad = 57.3°

1° = 0.0175 rad	60° = $\pi/3$ rad
15° = $\pi/12$ rad	90° = $\pi/2$ rad
30° = $\pi/6$ rad	180° = π rad
45° = $\pi/4$ rad	360° = 2π rad

1 rev/min = $\pi/30$ rad/s = 0.1047 rad/s

Speed

1 m/s = 3.60 km/h = 3.28 ft/s = 2.24 mi/h
1 km/h = 0.278 m/s = 0.621 mi/h = 0.911 ft/s
1 ft/s = 0.682 mi/h = 0.305 m/s = 1.10 km/h
1 mi/h = 1.467 ft/s = 1.609 km/h = 0.447 m/s
60 mi/h = 88 ft/s

Force

1 N = 0.225 lb
1 lb = 4.45 N
Equivalent weight of a mass of 1 kg on Earth's surface = 2.2 lb = 9.8 N

Pressure

1 Pa (N/m^2) = 1.45×10^{-4} lb/in^2 = 7.5×10^{-3} torr (mm Hg)
1 torr (mm Hg) = 133 Pa (N/m^2) = 0.02 lb/in^2
1 atm = 14.7 lb/in^2 = 1.013×10^{5} N/m^2 = 30 in. Hg = 76 cm Hg
1 lb/in^2 = 6.90×10^{5} Pa (N/m^2)
1 bar = 10^{5} Pa
1 millibar = 10^{2} Pa

Energy

1 J = 0.738 ft·lb = 0.239 cal = 9.48×10^{-4} Btu = 6.24×10^{18} eV
1 kcal = 4186 J = 3.968 Btu
1Btu = 1055 J = 778 ft·lb = 0.252 kcal
1 cal = 4.186 J = 3.97×10^{-3} Btu = 3.09 ft·lb
1 ft·lb = 1.36 J = 1.29×10^{-3} Btu
1 eV = 1.60×10^{-19} J
1 kWh = 3.6×10^{6} J

Power

1 W = 0.738 ft·lb/s = 1.34×10^{-3} hp = 3.41 Btu/h
1 ft·lb/s = 1.36 W = 1.82×10^{-3} hp
1 hp = 550 ft·lb/s = 745.7 W = 2545 Btu/h

Mass–Energy Equivalents

1 u = 1.66×10^{-27} kg ↔ 931.5 MeV
1 electron mass = 9.11×10^{-31} kg = 5.49×10^{-4} u ↔ 0.511 MeV
1 proton mass = 1.673×10^{-27} kg = 1.007 267 u ↔ 938.28 MeV
1 neutron mass = 1.675×10^{-27} kg = 1.008 665 u ↔ 939.57 MeV

Temperature

$T_F = \frac{9}{5} T_C + 32$
$T_C = \frac{5}{9} (T_F - 32)$
$T_K = T_C + 273$

cgs Force

1 dyne = 10^{-5} N = 2.25×10^{-6} lb

cgs Energy

1 erg = 10^{-7} J = 7.38×10^{-6} ft·lb

Applications of Physics in Technology

BRIAN K. SALTZER

Pearson Custom Publishing, Boston MA   A Pearson Education Company

Cover designed by Mary Louise Dorfner. Photos courtesy of PhotoDisc.

Printed in the United States of America

10 9 8 7 6 5 4 3

Please visit our web site at www.pearsoncustom.com

ISBN 0-536-62265-5

BA 993530

PEARSON CUSTOM PUBLISHING
75 Arlington Street, Suite 300, Boston, MA 02116
A Pearson Education Company

For my father, Richard Saltzer, with love and thanks.

Contents

Preface

The Purpose of the Book

For many students, a first course in physics is a test of endurance rather than a rewarding educational experience. Too often they merely plug numbers into equations and work problems without ever gaining a conceptual understanding of the material. While this is sad for any student, it is a doubly unfortunate for a student of technology.

Although it is not always obvious, the student of electronics or structural design is actually studying *physics;* only the language of the presentation differs. In this book, we will draw upon the primary concepts of physics introduced in Jerry D. Wilson and Anthony J. Buffa's *College Physics*, and show how they apply to technological situations that are meaningful to the student. These applications range from motivating Kirchhoff's Laws from the principles of energy, to the effects of thermal expansions on structures. Our hope is that by presenting the language and principles of physics in an interesting setting, the students will appreciate the applicability of physics. Instead of being a subject that the students simply endure, physics will become a tool they can use to give them a better understanding of the concepts related to their careers.

The Structure of the Book

In the two sections of the book, *Applications of Physics in Structural Analysis and Design* and *Applications of Physics in Electronics*, we apply the concepts of physics to structural analysis and design and to electronics. In these sections, the applications are written using either the students' chosen discipline as a setting, or using language and terms that they will encounter in their field. Because the purpose is to demonstrate real situations in which students can apply the tools of physics, we have tried to choose concepts of direct relevance to the field. Thus, different physical concepts are emphasized in the two sections. Although a thorough understanding of statics is beneficial to a drafter or structural designer, this topic is of secondary importance for a student of electronics. Conversely, the knowledge that Kirchhoff's Laws are rooted in the physical principle of conservation of energy, while beneficial to an electronics student, is not of primary importance for a computer drafter/designer.

To help students navigate between the Wilson/Buffa text and this Applications book, we have created connection boxes to guide the students to the appropriate section in *College Physics* or to another application that includes useful background information. At the end of each application, exercises are included to ensure that the students have absorbed the material discussed in the application. Because the curriculum calls for the exercises to be assigned as homework in addition to selected exercises from the *College Physics* text, the number of exercises is kept to a minimum.

Applications of Physics in Structural Analysis and Design

The first section of the book includes fifteen applications written to help reinforce many of the principles from *College Physics* that are most important to the students.

They are:

1. Unit Conversions and 3-D Design
2. Scalars, Vectors, and Design
3. Equal and Opposite Forces
4. Static Systems with More than One Mass
5. The Normal Force and Friction
6. Using a Rotated Coordinate System to Analyze a Design
7. Systems in Translational Equilibrium Along Both the X-and Y-Axes
8. Torque
9. Rotational and Translational Equilibrium: A System with Three Unknowns
10. The Effects of Thermal Expansions in Static Situations
11. Columns, Beams, and Trusses
12. Arches
13. Centripetal Force and the Inclination of a Banked Turn
14. Heating, Ventilation, Air-Conditioning, and the Continuity Equation
15. The Foucault Pendulum

The section opens with two preliminary applications that associate the language of unit conversions, scalars, and vectors with the language of design. We show the students how these preliminary physics principles apply to their chosen field of study.

After this introduction, the student moves on to Applications 3-10 that focus on the subject of statics. Because this topic is of critical importance to a design student, it is broken into eight smaller pieces that slowly guide the student through the subject. As the student works through each statics application, a new concept is added to the previously discussed material until the student is finally confronted with a full, three-dimensional static situation that allows for a discussion of thermal expansion and several simultaneous unknowns.

The final five applications focus on design variables and elements that are representative of those that the students may confront in their careers. The coverage ranges from a physical discussion of columns, beams, and trusses to an explanation of heating, ventilation, and air-conditioning in terms of the Continuity Equation from fluid dynamics.

Applications of Physics in Electronics

The second section of the text contains fifteen applications of physics to electronics. It is laid out in the same manner as *Applications of Physics to Structural Analysis and Design*, but focuses instead on those physical concepts that are of most interest to a student of electronics:

1. Significant Figures and the Charge on the Electron
2. Unit Analysis and Conversions in Electronics
3. Density and Electronics
4. Vectors and Electronics
5. The Force on an Electron in an Electric Field
6. The Force Between Two Current-Carrying Wires
7. A Comparison of the Gravitational and Electromagnetic Forces
8. Potential Versus Potential Energy
9. Applying the Principles of Energy to Series and Parallel Resistors
10. Heat, Kinetic Energy, and the Breakdown of PN-Junction Diodes
11. The Power Dissipated in an RLC- Circuit

12. The Electromagnetic Signal Produced by a Dipole Antenna
13. Diffraction and Television Signals
14. Relativistic Kinetic Energy and Momentum
15. Physics and the Next Generation of Computing

The focus of this section is on giving students a deeper understanding of the electronic principles with which they are familiar and to give them a *physical* introduction to a few of the concepts that they will be exposed to in subsequent studies. The topics covered range from an explanation of Kirchhoff's Laws using the principle of potential energy to an explanation of the radio signal that is produced by an antenna.

‣ Supplements

An instructor's Resource Manual containing solutions to the applications exercises and a third section to the book, *Applications of Calculus in Physics*, is available from the publisher. A companion website will also be available to support the book.

Acknowledgements

First, I would like to thank the members of the ITT Technical Institute curriculum team for their continuing dedication to quality education. This book was conceived as a result of their desire to make physics more interesting and rewarding for technical students. Thanks also goes to Frank Burrows and the production staff at Pearson Custom Publishing for making the book possible.

Next, I would like to thank the faculty members and students of ITT Technical Institute – Pittsburgh for their time and efforts in reviewing the manuscript and working all of the exercises. As always, it's a privilege to be associated with you.

I also owe considerable gratitude to the review team for this project – Harvey Rumbaugh (ITT Technical Institute – Pittsburgh), Alexander Devereux, Dilman Longbrake, Thaiyar Srinivasan (ITT Technical Institute – Phoenix), Patricia Battler (ITT Technical Institute – Tampa), Murray Bourne (Ngee Ann Polytechnic), Chester Benson (copyeditor), and Sarah Streett (accuracy checker for the calculus exercises). Thank you for your many helpful comments and suggestions. In particular, special thanks goes to John Sumner (University of Tampa) and Kevin Handel for their extremely detailed comments, suggestions, and corrections.

Lastly, my largest thanks goes to Ann Heath (editor) and Eric Stimmel (art/formatting). Thank you for your dedication, willingness to work insane hours, and your ability to work in a "nonlinear" fashion. I could not have done it without you.

Brian K. Saltzer
Pittsburgh, PA

Part I: Structural Analysis and Design

Courtesy of Corbis Digital Stock

This part contains fifteen applications that illustrate how the language and concepts of physics apply to the field of structural analysis and design.

The first two applications show how the preliminary physics principles of unit conversions, scalars, and vectors apply to problems in design.

Applications 3-10 focus on the critically important topic of statics. We build this topic slowly through eight applications that gradually increase in level of complexity. The culminating application explores a three-dimensional static situation involving thermal expansion and several simultaneous unknowns.

The final five applications explore several design variables and elements that are fundamental to the design problems we encounter in the workplace. They range from a physical discussion of columns and beams, like those found at Stonehenge, to an explanation of heating, ventilation, and air-conditioning in terms of the Continuity Equation from fluid dynamics.

Part I: Structural Analysis and Design

Applications

1. Unit Conversions and 3-D Design
2. Scalars, Vectors, and Design
3. Equal and Opposite Forces
4. Static Systems with More than One Mass
5. The Normal Force and Friction
6. Using a Rotated Coordinate System to Analyze a Design
7. Systems in Translational Equilibrium Along Both the X-and Y-Axes
8. Torque
9. Rotational and Translational Equilibrium: A System with Three Unknowns
10. The Effects of Thermal Expansions in Static Situations
11. Columns, Beams, and Trusses
12. Arches
13. Centripetal Force and the Inclination of a Banked Turn
14. Heating, Ventilation, Air-Conditioning and the Continuity Equation
15. The Foucault Pendulum

STRUCTURAL ANALYSIS AND DESIGN 1

Unit Conversions and 3-D Design

In this application, we employ the unit conversion techniques introduced in *College Physics* to examine a few of the common three-dimensional forms used in design.

The *units* attached to a measurement are labels that identify the type of quantity being measured. For example, the expression 5 seconds is a measurement of time while 5 inches is a measurement of length. Although both measurements have a numerical value of 5, it is the unit label of seconds or inches that identifies the physical quantity being measured.

Connection

Units of measure are introduced in Section 1.2 of *College Physics*.

As discussed in Section 1.2 of Wilson/Buffa's *College Physics*, a number of different systems of measurement with different units are used throughout the world. Because different people find different systems of units preferable when designing, it is important to be able to translate measurements from one system to another. This translation is called a *unit conversion*. It is important to note that when we execute a unit conversion, we do not change the size of the measurement, only the units in which it is being expressed.

A listing of common conversion factors can be found on the inside cover of *College Physics* or this text.

In this application, we will apply the methods of unit conversion discussed in Section 1.5 of *College Physics* to a sample of the three-dimensional shapes commonly used in design.

Using Conversion Factors

A *conversion factor* is a relationship such as

$$1\,\text{in} = 2.54\,\text{cm} \qquad (1)$$

that tells us that an object that is 1 inch long has the same length as an object that is 2.54 centimeters long. However, since we want to change the *units* on the measurement but not the *size* of the measurement, we must be careful how we apply the conversion factor to the measurement. Notice that we can take Equation (1) and divide both sides by 1 inch, leaving 1 on the left side of the expression:

$$1\text{ in} = 2.54\text{ cm}$$

$$\frac{1\text{ in}}{1\text{ in}} = \frac{2.54\text{ cm}}{1\text{ in}}$$

$$1 = \frac{2.54\text{ cm}}{1\text{ in}} \qquad \textbf{(2)}$$

Or, conversely, we can divide both sides of Equation (1) by 2.54 centimeters, leaving 1 on the right side of the expression:

$$1\text{ in} = 2.54\text{ cm}$$

$$\frac{1\text{ in}}{2.54\text{ cm}} = \frac{2.54\text{ cm}}{2.54\text{ cm}}$$

$$\frac{1\text{ in}}{2.54\text{ cm}} = 1 \qquad \textbf{(3)}$$

Because multiplying by 1 does not change the size of the measurement, the rewritten form of the conversion factor can now be used to change the units on the measurement.

Example 1

Express the length measurement of 6 inches in centimeters.

Solution Because we need to eliminate the unit of inches, we use the form of the conversion factor that places inches in the denominator so that inches will cancel when we change the units on the measurement.

$$6\text{ in} \cdot \frac{2.54\text{ cm}}{1\text{ in}}$$

Once the unit of inches has been cancelled, the only unit remaining in the expression is centimeters. By carrying out the required multiplication, we are successful in changing the units on the measurement from inches to centimeters.

$$6\text{ in} \cdot \frac{2.54\text{ cm}}{1\text{ in}} = 15.24\text{ cm}$$

> **FYI**
>
> Notice that because a centimeter is smaller than an inch, it takes more centimeters to express the same length.

We find that an object that is 6 inches long has the same length as one that is 15.24 centimeters long.

Three-Dimensional Unit Conversions

In three-dimensional design, the units attached to a measurement are slightly more complicated than those used in a one- or two-dimensional situation. For example, the units for the volume of a cube are found by multiplying the units attached to each side of the cube.

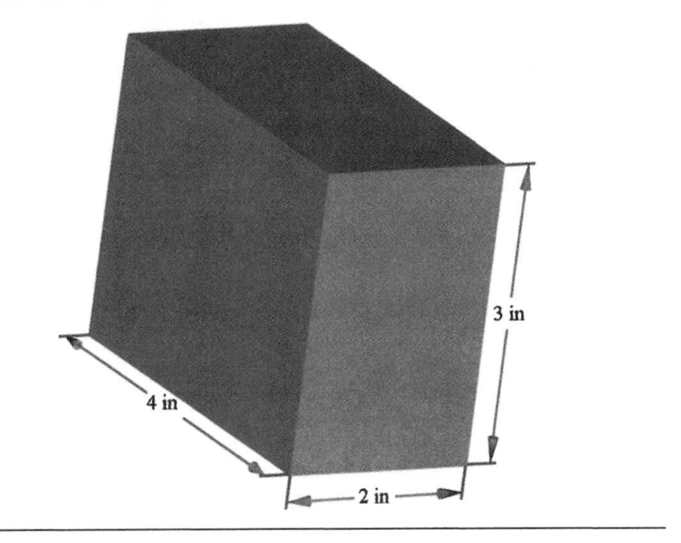

Figure 1.1

$$\text{Volume} = (2\text{ in})(3\text{ in})(4\text{ in})$$
$$\text{Volume} = (2)(3)(4)\text{in} \cdot \text{in} \cdot \text{in}$$
$$\text{Volume} = 24\text{ in}^3$$

To express the volume of this cube in the units of cubic centimeters, we will use Equation (2) because we need to eliminate the unit of inches. However, notice that inches in the denominator of the conversion factor can only cancel *one* of the inches in the volume units of the cube.

$$\text{Volume} = 24\text{ in}^3 \cdot \frac{2.54\text{ cm}}{1\text{ in}}$$
$$\text{Volume} = 24\text{ in} \cdot \text{in} \cdot \text{in} \cdot \frac{2.54\text{ cm}}{1\text{ in}}$$

Consequently, we need *three* identical copies of the conversion factor to cancel the units of cubic inches.

$$\text{Volume} = 24\,\text{in}\cdot\text{in}\cdot\text{in}\cdot\frac{2.54\,\text{cm}}{1\,\text{in}}\cdot\frac{2.54\,\text{cm}}{1\,\text{in}}\cdot\frac{2.54\,\text{cm}}{1\,\text{in}} \quad \textbf{(4)}$$

An expression such as Equation (4) is often written in the more compact form

$$\text{Volume} = 24\,\text{in}\cdot\text{in}\cdot\text{in}\left(\frac{2.54\,\text{cm}}{1\,\text{in}}\right)^3$$

Executing the required multiplication and canceling the unit of inches

$$\text{Volume} = 393.29\,\text{cm}\cdot\text{cm}\cdot\text{cm}$$

$$\text{Volume} = 393.29\,\text{cm}^3$$

results in the discovery that a cube with a volume of 24 cubic inches also has a volume of 393.29 cubic centimeters.

Example 2

Given a sphere with a radius of 6 cm,

a) calculate the sphere's volume in cubic centimeters
b) convert the result from cubic centimeters into cubic inches

Solution The volume of a sphere is found using the equation

$$\text{Volume} = \frac{4}{3}\pi r^3$$

Inserting the radius of the sphere into the equation gives us the volume expressed in cubic centimeters:

$$\text{Volume} = \frac{4}{3}\pi r^3$$

$$\text{Volume} = \frac{4}{3}\pi(6\,\text{cm})^3$$

$$\text{Volume} = 904.78\,\text{cm}^3$$

We will now use Equation (3) to convert the units of volume to cubic inches

$$\text{Volume} = 904.78\,\text{cm}^3 \cdot \left(\frac{1\,\text{in}}{2.54\,\text{cm}}\right)^3$$

$$\text{Volume} = 55.21\,\text{in}^3$$

Conclusion

The ability to convert units is an important skill for anyone wishing to enter the design field. Because of the variety of systems of units available, today's drafter/designer must be able to convert the units on a drawing or design in order to interact with other members of the design team. Also, today's designer must be able to work and communicate with designers of many different nationalities. Because the United States uses the British system of units, while much of the world uses the metric system, a successful drafter/designer must be able to move between these two systems of measurement.

Exercises

1. A certain sphere has a radius of 5 cm.

   a) Calculate the volume of the sphere in cm^3
   b) Convert the volume into in^3
   c) Convert the volume into ft^3

2. Calculate the volume of a cube that is 8 inches on each side. Express the volume in both ft^3 and cm^3.

3. The volume of a cone that has a base radius of r and a height h is given by

$$V = \frac{1}{3}\pi r^2 h$$

   For the cone that has a base radius of 0.5 ft and a height of 2 feet,

   a) calculate the volume of the cone in ft^3
   b) convert your answer from a) into in^3
   c) convert your answer from b) into cm^3

4. The *frustrum of a cone* is a cone that has been truncated and thus has an upper radius, r_1, and a base radius, r_2.

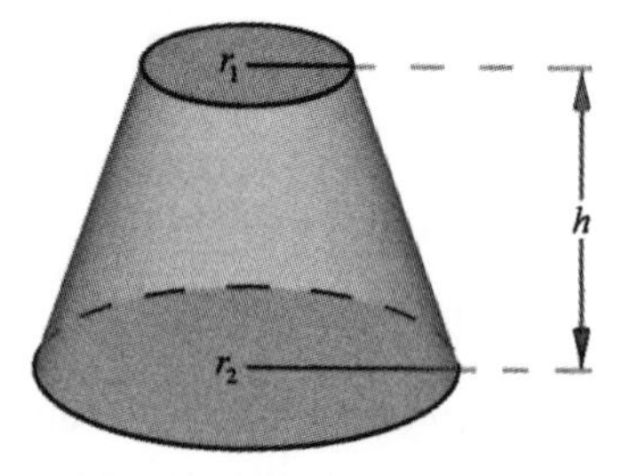

The volume for the illustrated frustrum is found using

$$V = \frac{1}{3}\pi h\left(r_1^{\,2} + r_1 r_2 + r_2^{\,2}\right)$$

For the frustrum that has an upper radius of 5 mm, a base radius of 10 mm, and a height of 20 mm,

a) calculate the volume of the frustrum in mm^3
b) convert the answer from a) into cm^3
c) convert the answer from b) into in^3

Structural Analysis and Design 2

Scalars, Vectors, and Design

In this application, we use examples from structural design to illustrate the difference between vector and scalar quantities.

Connection

Scalars and Vectors are discussed in Sections 2.1 and 2.2 of *College Physics.*

Mass and Weight are discussed in Section 1.2 of *College Physics.*

Although many different physical quantities are involved in the design of a structure, it is possible to categorize all of them into one of two classifications. All physical quantities are classified as being either *scalars* or *vectors*. In this brief application, we will define these two terms, and use them to classify some of the more common physical quantities used in design.

A *scalar* quantity is a physical quantity that has a size (or magnitude), but not an associated direction. For example, the mass of a beam is a measurement of the amount of matter contained in the beam. Because mass does not have an associated direction, only a magnitude, it is a scalar quantity.

If we now consider the *weight* of the beam, we see that it has both a size and a direction (downward) and is therefore classified as a *vector* quantity.

Classifying Density

In Section 1.4 of *College Physics*, the density of a material is defined as being the amount of mass that is contained in a certain volume and is given by the equation

$$\rho = \frac{m}{V}$$

Because of the differences in the organization of the various molecules that comprise such materials as structural steel, concrete, etc., the amount of mass contained in a unit volume of each will differ. For example, structural steel has a different density than does concrete. However, the actual density of a material does not affect the classification. Because neither mass nor volume has an associated direction, the density of a material is classified as a scalar quantity.

Classifying Force

Consider the case of a beam like the one illustrated.

Figure 2.1

If we are interested in the response of the beam to an applied force, we need to know whether the force is applied from the left or from the right.

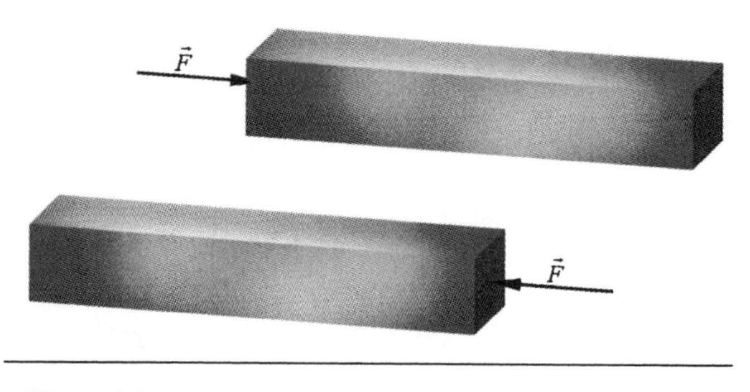

Figure 2.2

Because we must know the *direction* of the applied force in addition to the size of the force to discuss the response of the beam, force is classified as a *vector* quantity.

Classifying Temperature

Connection

Thermal Expansion is discussed in Section 10.4 of *College Physics.*

When designing structures such as bridges and highways, the engineer must be careful to allow for the span of temperatures that the structure will experience. For example, think about how sidewalks are generally constructed. Because of the expansion that concrete experiences at warmer temperatures, spaces are left in the sidewalk. These spaces allow for the expansion and contraction of the material between summer and winter, and day and night.

Several different scales of measure for temperature are discussed in the *College Physics* text. Regardless of whether the measurement is made in degrees Fahrenheit or Celsius, because the temperature of a material has a magnitude, but not an associated direction, temperature is a scalar quantity.

Conclusion

Although this application is a simple one which focuses exclusively on how to differentiate between scalars and vectors, the knowledge of whether we are working with one quantity or the other is of critical importance when we begin to analyze the physical properties in structural analysis and design.

Exercises

1. Discuss the difference between a vector quantity and a scalar quantity.

2. Carry out the necessary research to find three *physical* quantities used in each of the following areas of drafting/design.

   a) Civil drafting/design
   b) Architectural drafting/design
   c) Process piping
   d) Electronics drafting

3. For each of the physical quantities listed in Exercise 2, identify them as being either a scalar or a vector quantity.

STRUCTURAL ANALYSIS AND DESIGN 3

Equal and Opposite Forces

This application is the first in a series of examples that explore non-moving, or *static*, situations. We begin this series by studying a few examples that involve a single, static mass.

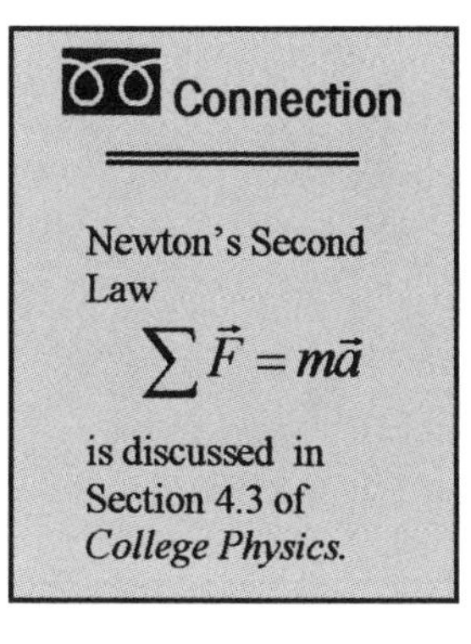

Newton's Second Law tells us that if a net force acts on a mass, the mass will accelerate in the direction of the applied force. Thus, if there is no net force operating on a mass, the mass will not accelerate.

Forces Acting on a Static Mass

Suppose that we have the following two forces acting on the mass *m:*

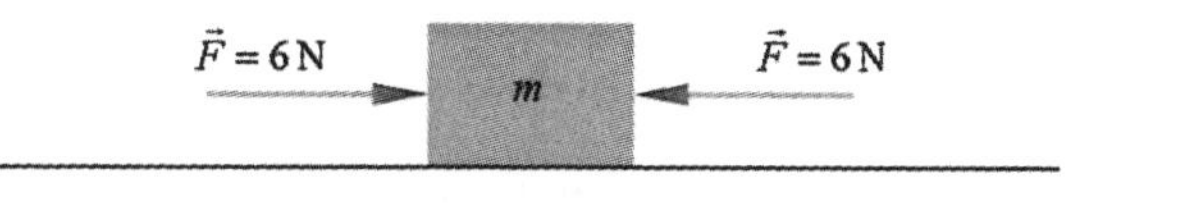

Figure 3.1

Review

In engineering and technology the term *static* means *no movement*. For example:

- An object is *static* if it is not moving;
- *static* friction is the friction that prevents an object from moving.

Because both force vectors have the same magnitude, 6 N, the mass will not move to the left or right. This is an example of a *static* situation. This static situation can also be expressed mathematically using Newton's Second Law:

$$\sum \vec{F} = 6\,\text{N} - 6\,\text{N} = 0$$

Example 1

What magnitude of $\vec{F}_2$ is necessary to prevent the mass from moving to the left?

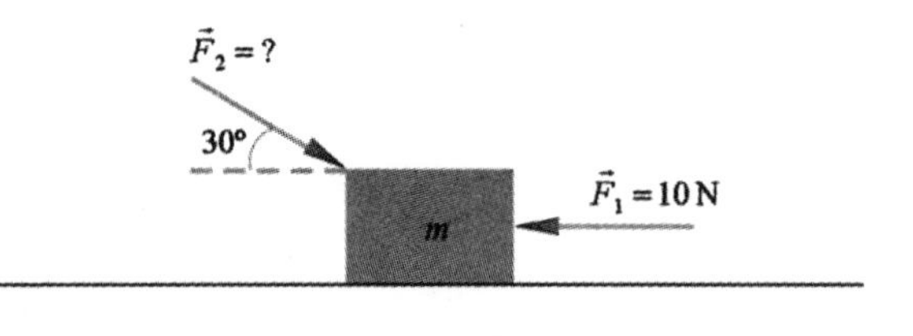

Figure 3.2

> **Connection**
>
> For a refresher on how to break a vector into its components, see Section 3.2 of *College Physics.*

Solution From our study of vector analysis, we know that only the horizontal component of $\vec{F}_2$ will act to oppose $\vec{F}_1$. Thus, to prevent the mass from moving horizontally we find

$$F_2 \cos 30° = F_1 \qquad \textbf{(1)}$$

Inserting the value for F_1 and solving for F_2 gives us

$$F_2 \cos 30° = F_1$$
$$F_2 \cos 30° = 10\ \text{N}$$
$$F_2(0.866) = 10\ \text{N}$$
$$F_2 = \frac{10\ \text{N}}{0.866}$$
$$F_2 = 11.5\ \text{N}$$

Example 2

Using the given figure, calculate the tension in the cable.

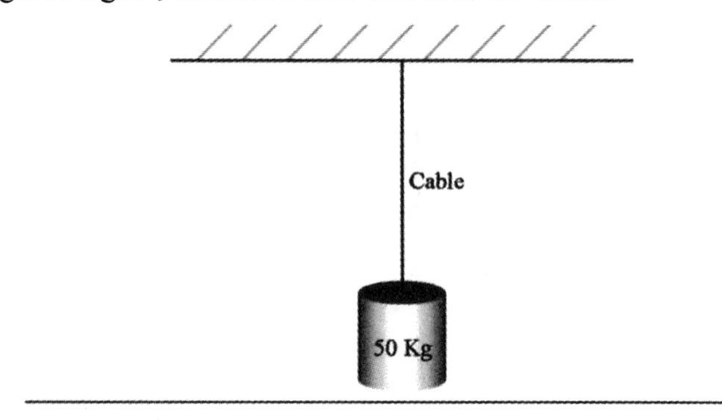

Figure 3.3

Solution As a first step, we will need to insert the forces acting on the mass.

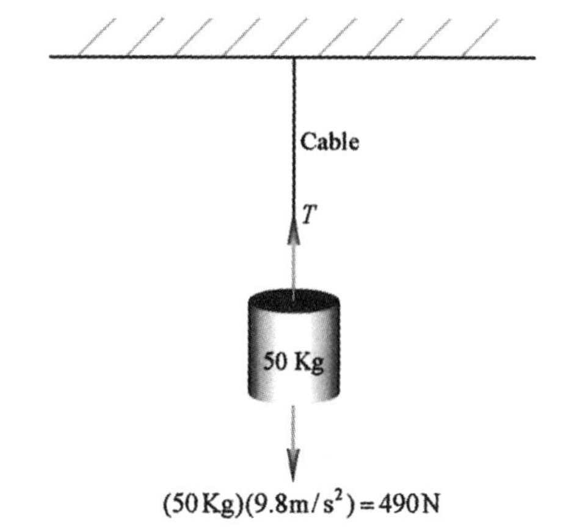

FYI

Notice that both the tension in the cable and the weight of the mass have directions and are therefore vectors. We will take these directions into account later in this application.

Figure 3.4

The weight of the mass is found by multiplying the mass by the gravitational acceleration.

$$W = mg$$
$$W = (50\ \text{Kg})(9.8\ \text{m/s}^2)$$
$$W = 490\ \text{N}$$

Because the mass is not moving upward or downward, the tension in the cable must be equal to the weight of the mass. Thus,

$$T = 490\ \text{N}$$

Examining Forces with Newton's Second Law

We can frame each of these examples in a little more formal language by using Newton's Second Law.

Revisiting Example 1

In the first example, because the mass was not moving to the right or to the left, Newton's law

$$\sum F_x = ma$$

becomes

$$\sum F_x = 0$$

If we choose the positive x-direction to point toward the right, Newton's law becomes

$$F_2 \cos 30° - F_1 = 0$$

which reproduces Equation (1)

$$F_2 \cos 30° = F_1$$

Revisiting Example 2

We can approach the second example in a similar way. Newton's Second Law applied along the vertical direction yields

$$\sum F_y = 0$$

Inserting the forces

$$T - 490 \text{ N} = 0$$
$$T = 490 \text{ N}$$

produces the same result that that we found using physical arguments.

Conclusion

One of the challenges faced by a designer is knowing how to plan for the various forces that will act on the structure or object. Knowing that an object or system is static provides a powerful tool for analysis. If an object, whether it be a beam, strut, arch, column, or some other structure, is not moving along a particular axis, we know that the forces along this axis must sum to zero. This realization, along with our knowledge of Newton's Second Law, often allows us to solve for forces and other unknowns in the system.

Exercises

1. What force, $\vec{F}_3$, would be required to keep the mass from moving?

$\vec{F}_1 = 7\text{ N}$ $\vec{F}_2 = 5\text{ N}$ m $\vec{F}_3 = ?$

2. What force, $\vec{F}_2$, would be required to keep the mass from moving?

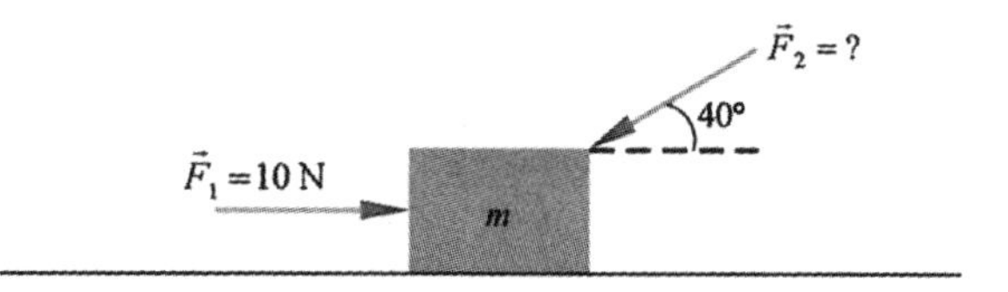

3. At what angle above the horizontal must $\vec{F}_2$ be placed in order to keep the mass from moving?

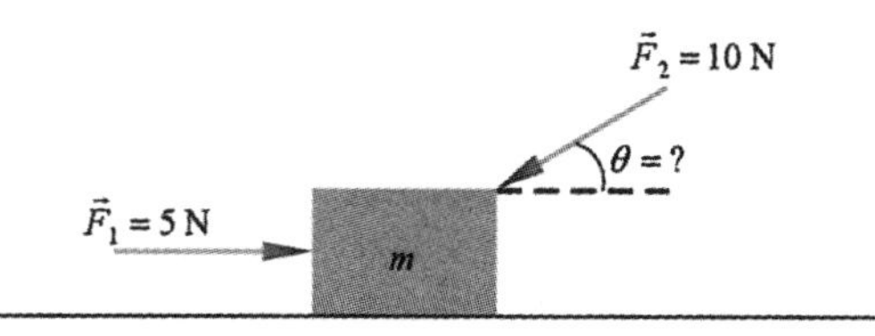

4. Calculate the tension in the cable.

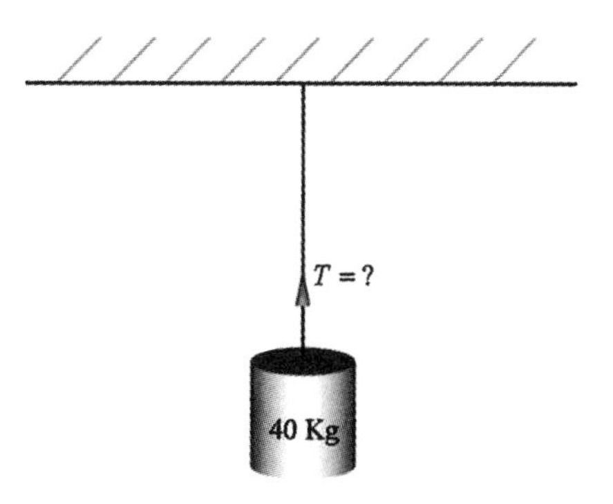

5. What force, $\vec{F}_3$, is required to keep the mass from moving?

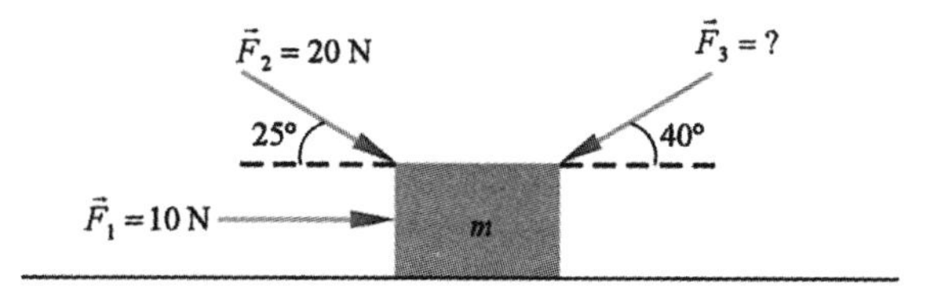

STRUCTURAL ANALYSIS AND DESIGN 4

Static Systems with More than One Mass

This application extends our use of Newton's Second Law in static situations to systems with more than one mass or section.

> **Connection**
>
> It will be helpful to work Structural Analysis and Design 3 - *Equal and Opposite Forces,* before attempting this application.

In Application 3, we analyzed the forces that were applied to a single stationary mass. In each example, we demonstrated that because the mass was static, the forces being applied totaled to zero. In this application, we will extend our use of Newton's Second Law to systems that have more than one mass or section. The purpose is to illustrate that when analyzing a system, it is only necessary to focus on the forces that act directly on each section.

Using Newton's Second Law

Find the tension in each cable of the given figure.

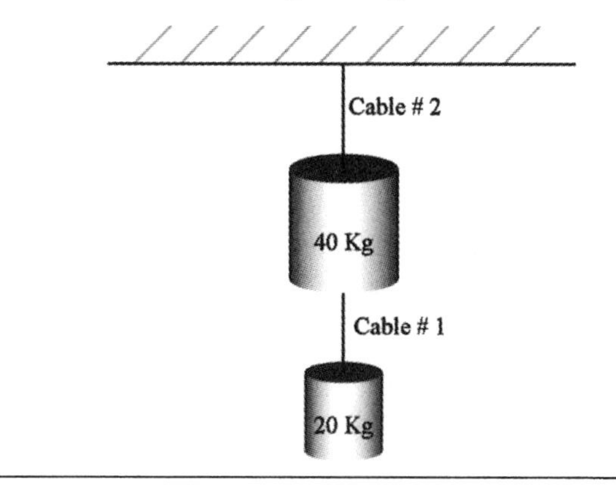

Figure 4.1

First, we insert all of the forces that act on the two masses.

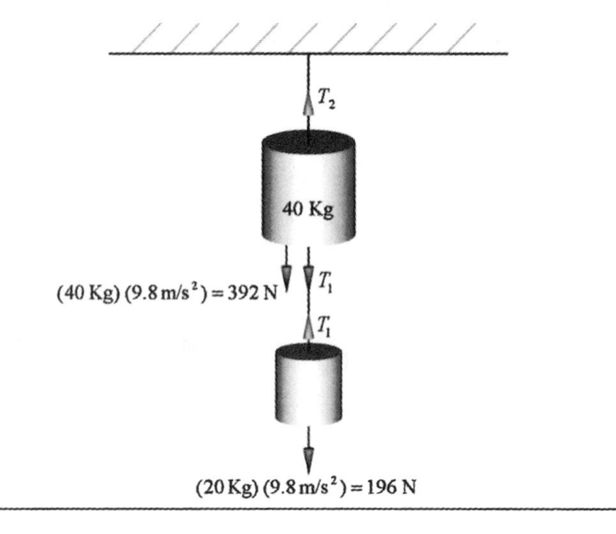

Figure 4.2

To apply Newton's Second Law to each mass, we only need to address the forces acting directly on each:

20 Kg:

$$\sum F_y = 0$$
$$T_1 - 196\text{ N} = 0$$
$$T_1 = 196\text{ N}$$

40 Kg:

$$\sum F_y = 0$$
$$T_2 - T_1 - 392\text{ N} = 0$$
$$T_2 = T_1 + 392\text{ N}$$
$$T_2 = 196\text{ N} + 392\text{ N}$$
$$T_2 = 588\text{ N}$$

In concluding this tension problem, we can argue intuitively that cable #1 only needs to support the weight of the 20 Kg mass (196 N), while cable #2 must support the weight of both masses (588 N).

The Force on a Truss

Trusses are discussed in more depth in Structural Analysis and Design Application 11.

A truss is a series of struts that are arranged in triangular patterns as shown in the figure. This structure is often used to construct bridges.

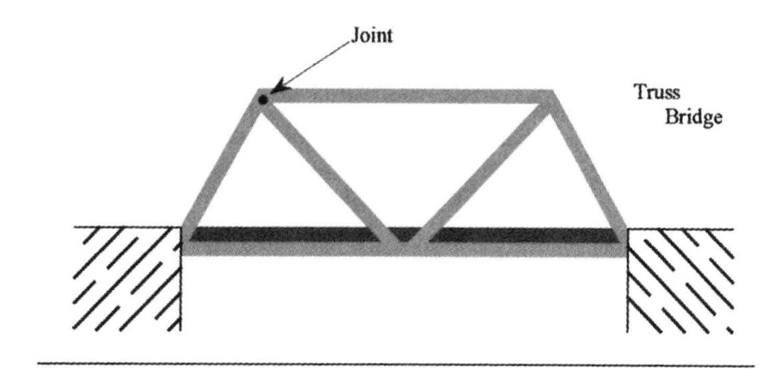

Figure 4.3

Because each strut in the structure is static, we have the freedom to analyze each one independently. Since each strut is not moving upward or downward, the forces along both the x- and y-axes must total to zero *for each strut*. Therefore, it is possible to analyze the entire truss as a single static whole, or to look at each piece individually during our design analysis.

Conclusion

The technique discussed in this application is an extremely powerful tool in the analysis of static situations. Because we only need to address those forces that act directly on an object, we can take a complex system with many different forces and sections and break it into smaller, simpler pieces. Knowing that the entire system is static tells us that the forces acting directly on *each section* of the system must also total zero.

Exercises

1. Find the tension in cable #1 and cable #2.

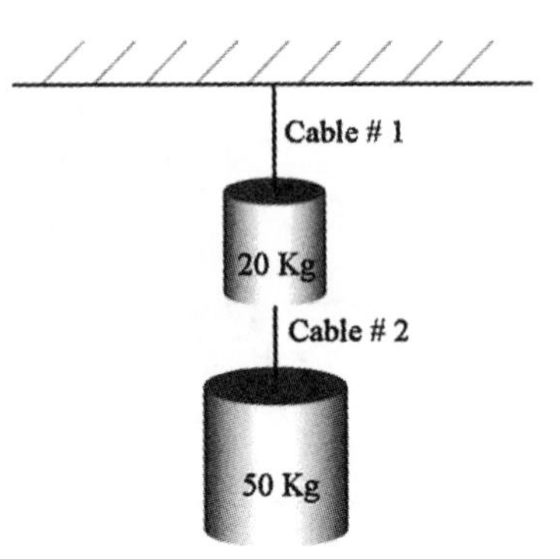

2. Find the tension in each cable.

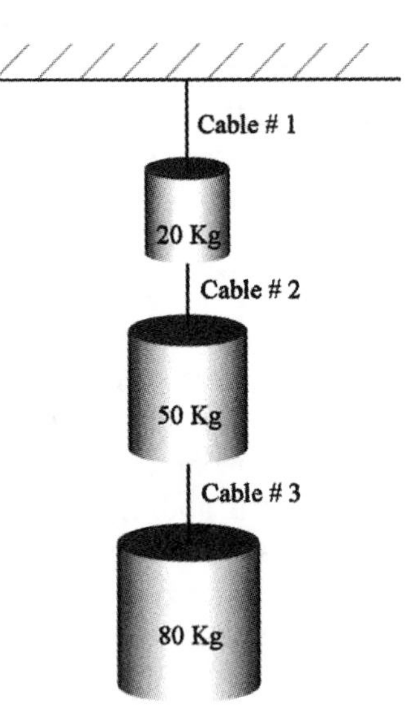

3. Find the tension in each cable. Hint: Begin by analyzing the forces acting on the knot at P.

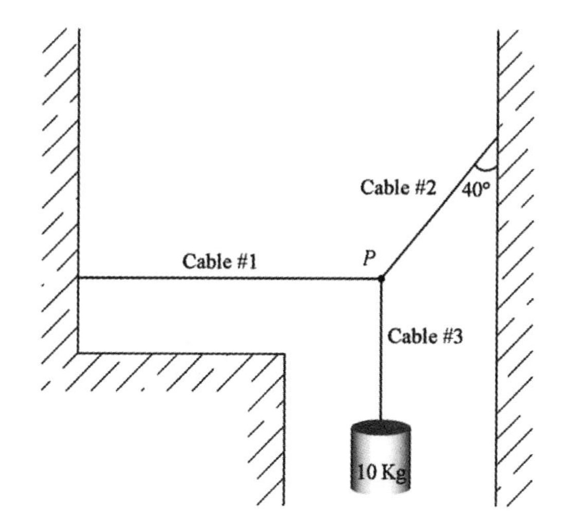

4.

a) What mass, m, is required to generate a tension of 490 N in cable #1?

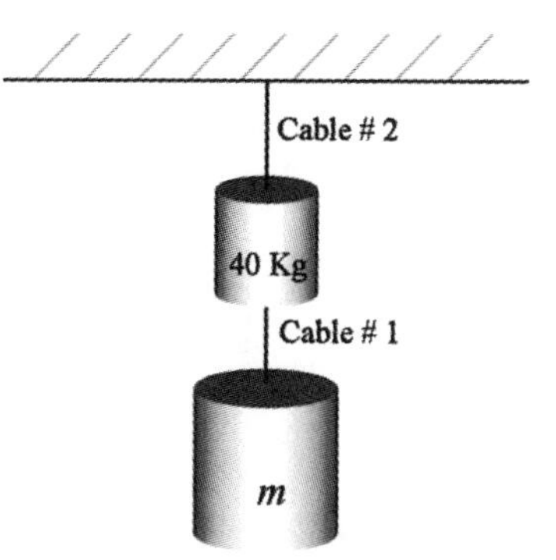

b) Using the fact that the tension in cable #1 is 490 N, find the tension in cable #2.

STRUCTURAL ANALYSIS AND DESIGN 5

The Normal Force and Friction

In this application, we use Newton's Second Law to motivate the existence of the normal force exerted by a surface and its relationship to friction.

Connection

For more information about the normal force, see Section 4.4 of *College Physics*.

In the two previous statics applications, we used Newton's Second Law to analyze static objects and systems. Because the masses were not moving, we argued that

$$\sum \vec{F} = m\vec{a}$$

can be simplified to

$$\sum \vec{F} = 0$$

We can now use the fact that the forces sum to zero for a static object to motivate the existence of another force being exerted on the object, the *normal force*.

The Normal Force

Suppose that a mass is resting on a horizontal surface like that shown in the figure.

Figure 5.1

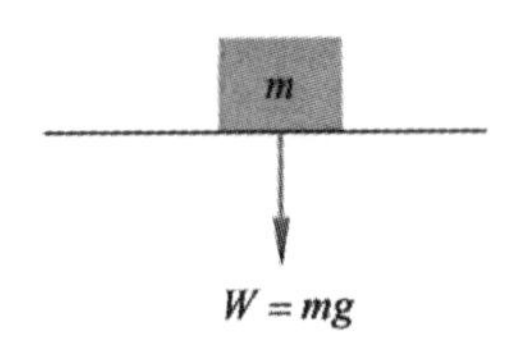

Figure 5.2

Our previous studies have demonstrated that because the mass is static (it is not moving upward or downward), the forces along the vertical axis must add up to zero. For this to be true, the weight vector cannot be the only force acting on the mass. An upward-directed force exerted by the surface on the mass must be present to counter the downward-directed weight.

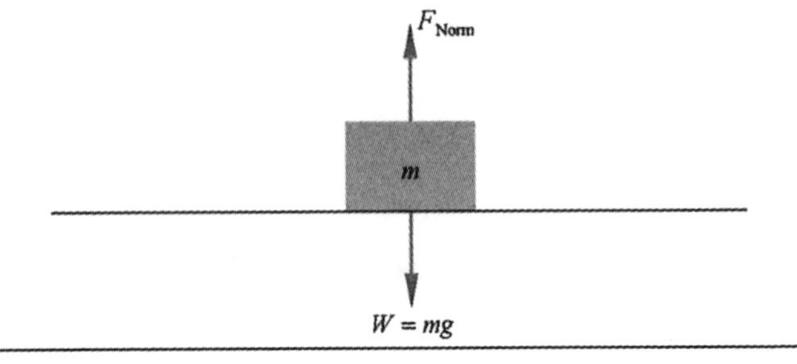

Figure 5.3

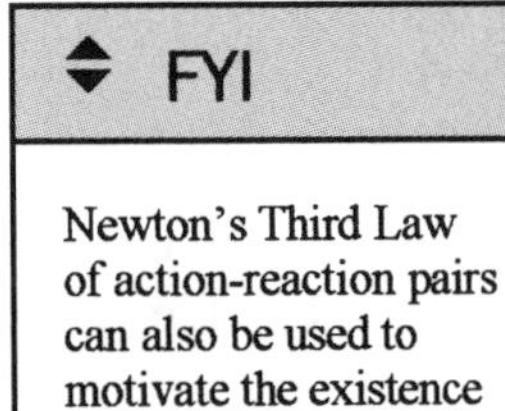

This upward-directed force that counteracts the weight of the mass is called the *normal force*.

Characteristics of the Normal Force

1. The normal force acts to counter all of the inward-directed forces exerted on a surface.
2. The normal force is always oriented perpendicular to the surface.

Example 1

Find the normal force exerted by the surface on the mass.

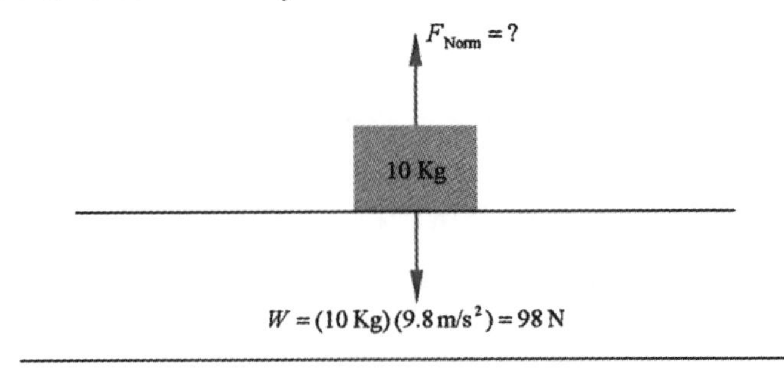

Figure 5.4

Solution As we discussed previously, the normal force must counter all of the force exerted on the surface. In this case, the amount of downward-directed force is simply the weight of the mass. This weight can be found by multiplying the mass by the gravitational acceleration, yielding a weight of 98 N.

Therefore, the surface must provide an upward-directed force of 98 N to counter the weight of the mass.

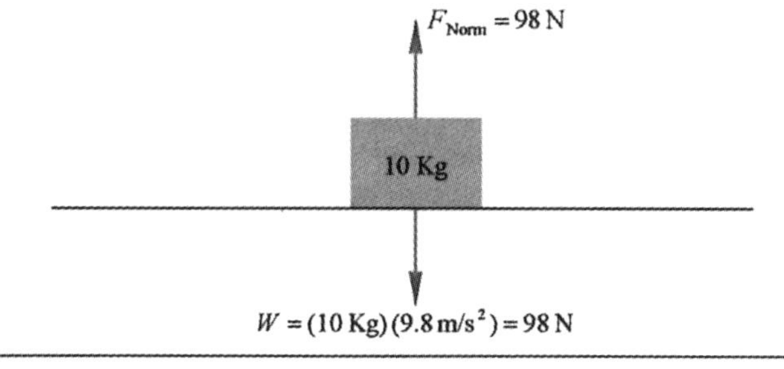

Figure 5.5

Example 2

Find the normal force exerted by the surface on the mass.

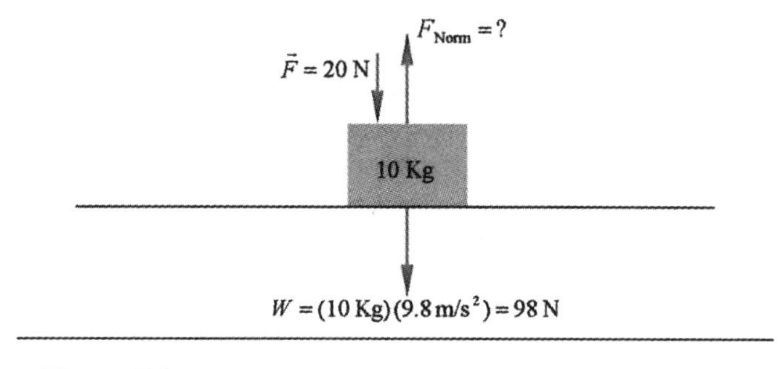

Figure 5.6

Solution In this example, there are two downward directed forces, the weight of the 10 Kg mass and an additional 20 N force exerted downward on the mass. To support the mass and keep it static, the surface must exert an upward-directed normal force equal in magnitude to all of the downward-directed force. Thus,

$$F_{\text{Norm}} = 20\text{ N} + 98\text{ N}$$
$$F_{\text{Norm}} = 118\text{ N}$$

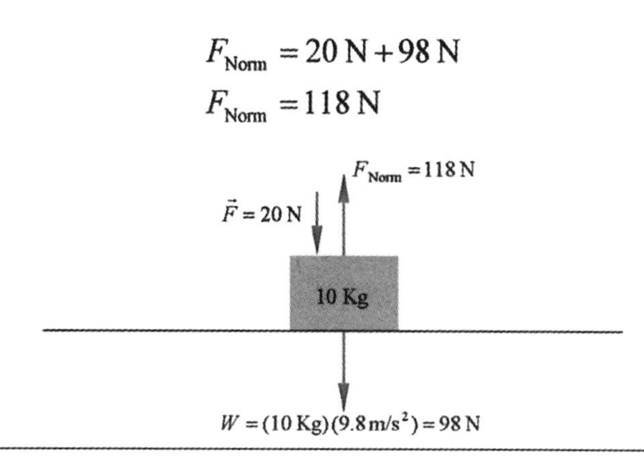

Figure 5.7

Example 3

Find the normal force exerted by the inclined plane on the 10 Kg mass.

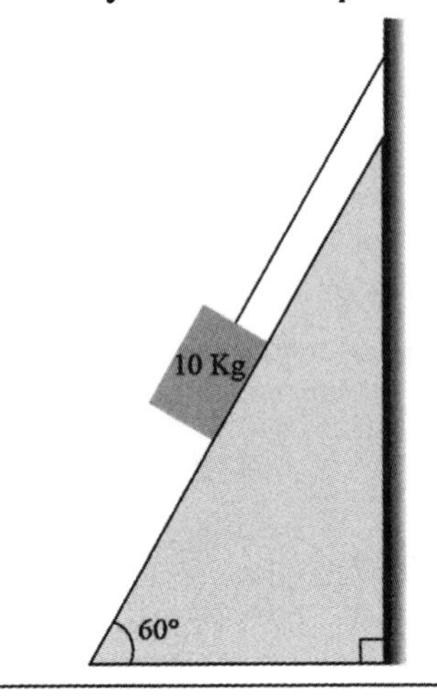

Figure 5.8

Solution This example is slightly more complex because of the 60° incline of the plane and the orientation of both the weight vector and the normal force. Because the normal force is always exerted perpendicular to the surface, it is not vertical in this example but is instead oriented at an angle:

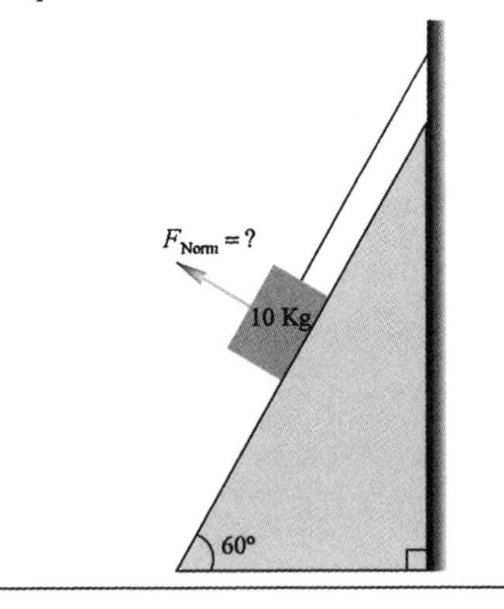

Figure 5.9

Connection

For a refresher on how to break a vector into components, see Section 3.2 of *College Physics*.

Recall that the normal force is the reaction of the surface to the total amount of force exerted perpendicular to the surface. Therefore, the normal force is not simply equal to the weight of the mass. In this example, the normal force only needs to counter the amount of the weight vector that is directed inward on the surface.

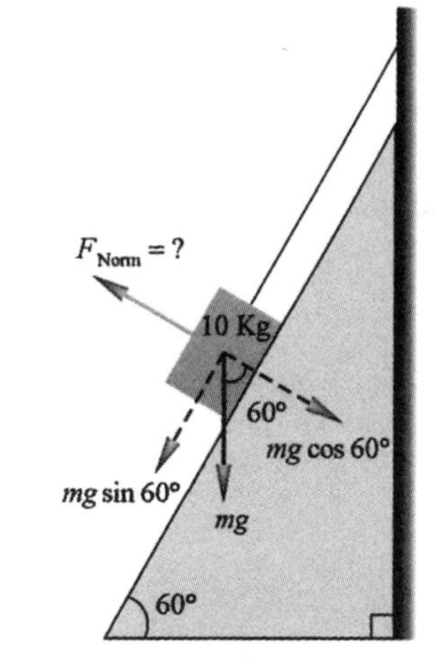

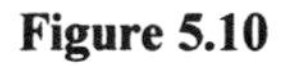
Figure 5.10

Thus,

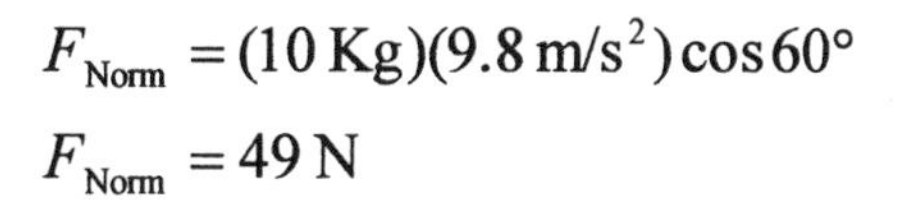

$$F_{Norm} = (10\ \text{Kg})(9.8\ \text{m/s}^2)\cos 60°$$
$$F_{Norm} = 49\ \text{N}$$

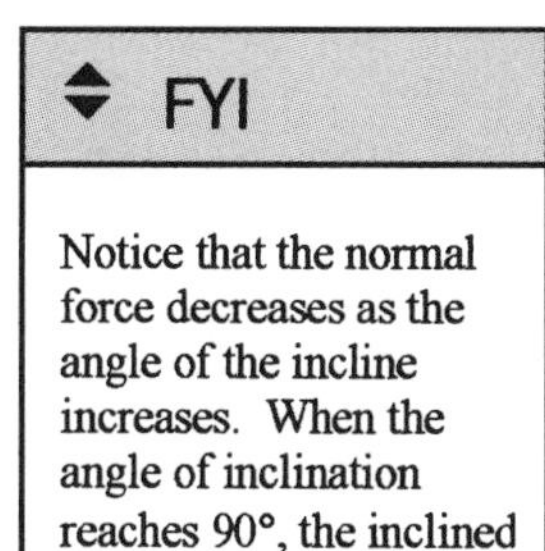

FYI

Notice that the normal force decreases as the angle of the incline increases. When the angle of inclination reaches 90°, the inclined plane will not exert any normal force on the mass.

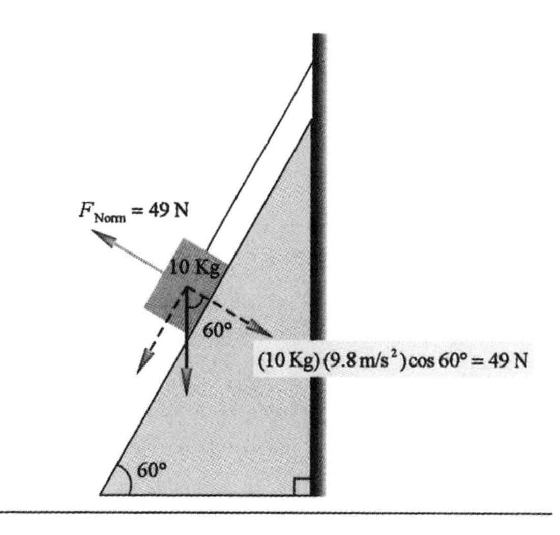

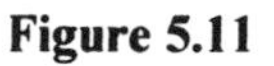
Figure 5.11

Friction

Up to this point in the development, we have examined the effect of the normal force on objects, but we have failed to take into account another important real-world force, *friction*. To consider more realistic examples, we must add friction to our analysis.

Friction is found in two forms, *kinetic friction* and *static friction*. Kinetic friction is the force that an object must continue to overcome as it slides along a surface. Kinetic friction is often expressed as f_K.

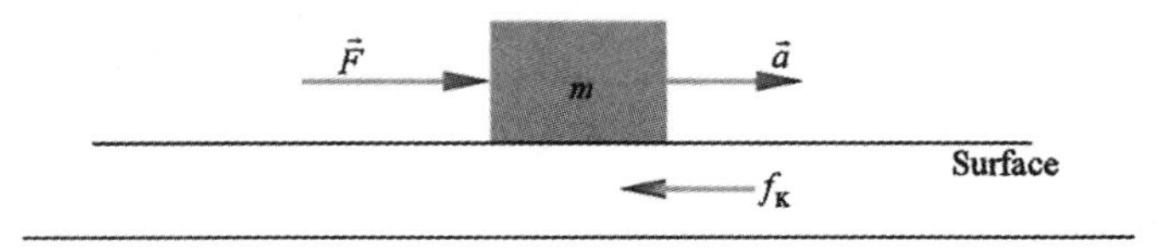

Figure 5.12

Static friction is the force between the object and the surface that prevents the object from moving along the surface. Static friction is often expressed as f_S.

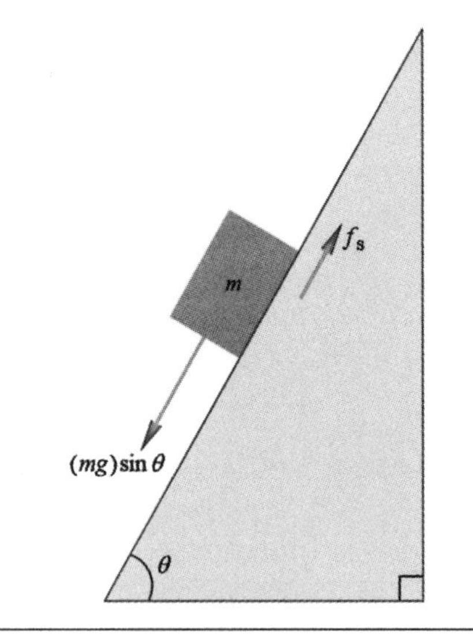

Figure 5.13

Because these two types of friction are forces, they can be inserted into the sum of forces in Newton's Second Law as we analyze a physical situation.

Coefficients of Friction

If we slide two objects past one another, the physical characteristics of each object will come into play. For example, it is much easier to slide two panes of glass, separated by a thin film of water, past one another than two pieces of rubber. Information about the two types of objects is contained in a quantity known as the *coefficient of friction* (μ). This coefficient is defined as the ratio of the amount of frictional force on an object to the size of the normal force on the object.

$$\mu = \frac{f_{\text{Friction}}}{F_{\text{Norm}}}$$

It is interesting to note that because both the numerator and the denominator in this definition are measured in units of force, the units will cancel each other, leaving the coefficient of friction as a dimensionless quantity.

Just as there are two types of frictional forces, there are two types of coefficients of friction, kinetic and static.

Kinetic Friction: $$\mu_{\text{K}} = \frac{f_{\text{K}}}{F_{\text{Norm}}}$$

Static Friction: $$\mu_{\text{S}} = \frac{f_{\text{S}}}{F_{\text{Norm}}}$$

Both coefficient equations provide a simple way to find the actual frictional force exerted on an object. By multiplying both sides of each equation by F_{Norm}, we are left with the equation

$$f = \mu F_{\text{Norm}}$$

Using this equation, it is easy to find the amount of frictional force on an object by multiplying the coefficient of friction between the object and the surface and the normal force exerted on the object. The following example demonstrates this technique.

Example 4
Calculate the force of kinetic friction between the mass and the surface.

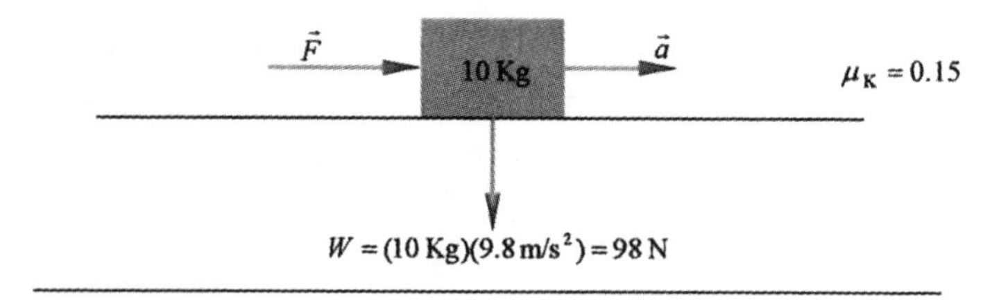

Figure 5.14

Solution Because the weight of the object is 98 N, we know that it also must be the size of the normal force exerted upward on the object by the surface. Consequently, we may find the solution by multiplying this normal force by the coefficient of kinetic friction given in the problem:

$$f_K = \mu_K F_{Norm}$$
$$f_K = (0.15)(98\text{ N})$$
$$f_K = 14.7\text{ N}$$

Example 5
What is the maximum value of tension that can be applied on the rope before the mass begins to move?

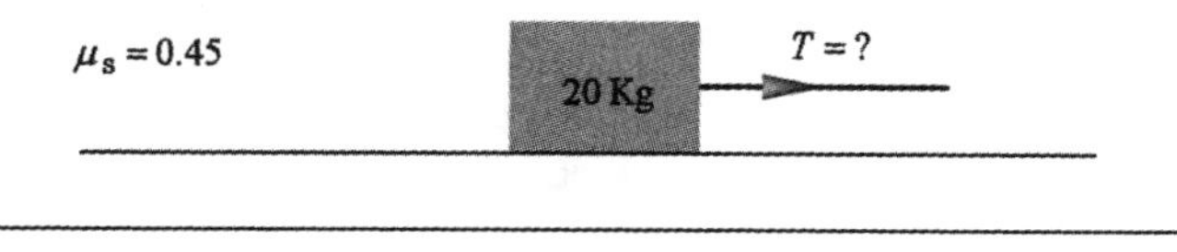

Figure 5.15

Solution The mass will move if the tension in the rope exceeds the force of static friction between the mass and the surface. Therefore, if we find this force of friction, we will have the solution.

The normal force in this case must be equal to the weight of the mass. Thus,

$$F_{Norm} = mg$$
$$F_{Norm} = 196\text{ N}$$

The force of static friction can now be found by multiplying the normal force by the coefficient of static friction given in the problem:

$$f_S = (0.45)(196\ \text{N})$$
$$f_S = 88.2\ \text{N}$$

Thus, if the tension in the rope exceeds 88.2 N, the mass will move.

Conclusion

The frictional force that exists between the surfaces of a structure, or other types of physical designs, must be taken into account in all aspects of the design. In order for a designer to choose the correct angles for the elements of the structure, the types of materials required, and many other facets of the design, the frictional forces exerted between the design elements must be known.

Exercises

1. Calculate the weight of a 100 Kg mass on earth.

2. Calculate the component of the weight of the mass that is perpendicular to the plane and the component of the weight that is directed down the plane.

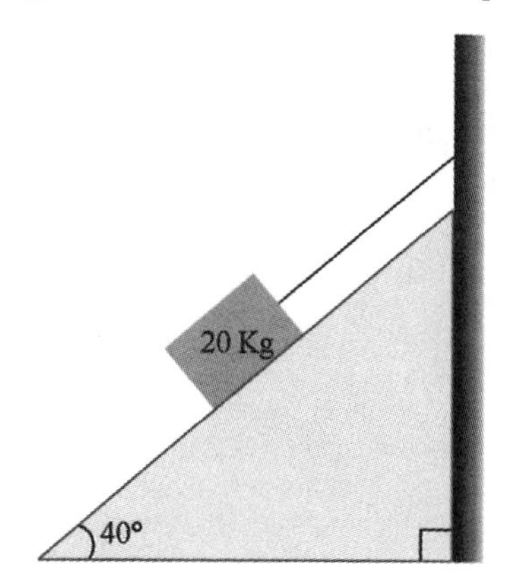

3. A 20 Kg mass is resting on a horizontal surface as in the following figure.

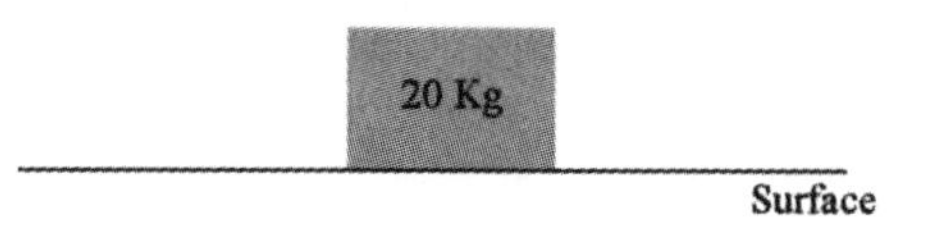

a) Calculate the weight of the mass.
b) What is the magnitude of the normal force exerted by the surface on the mass?

4.

a) What is the component of the weight of the mass perpendicular to the plane?
b) What is the magnitude of the normal force exerted by the plane on the mass?
c) Sketch the orientation of the normal force into the diagram.

5. Calculate the normal force exerted by the surface on the mass.

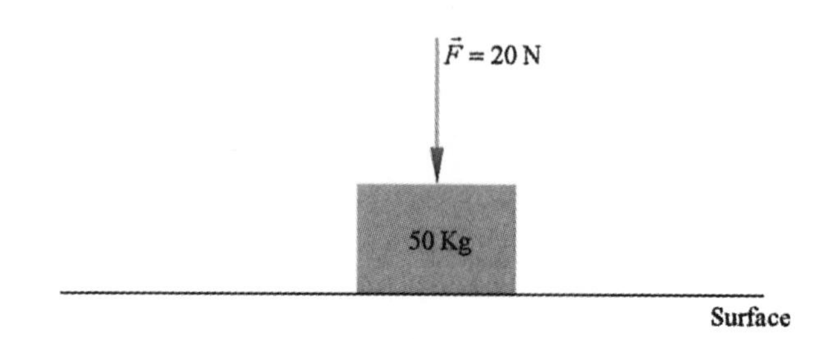

6. Calculate the normal force exerted by the surface on the mass.

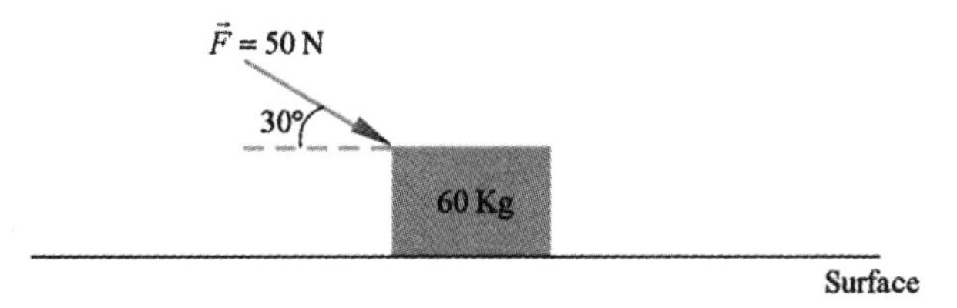

7. Calculate the normal force exerted by the inclined plane on the mass.

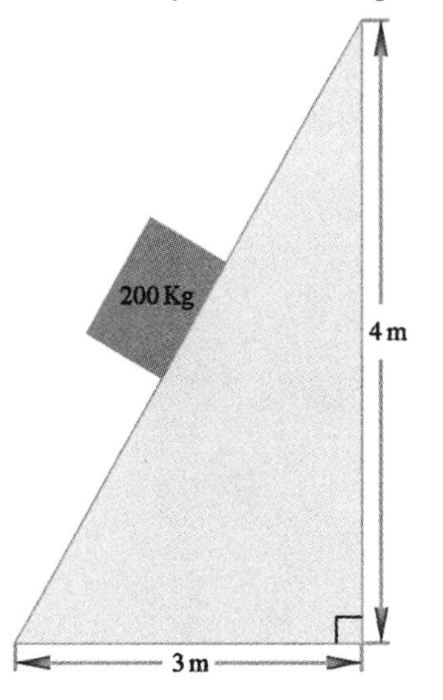

8. Calculate the force of kinetic friction between the mass and the surface.

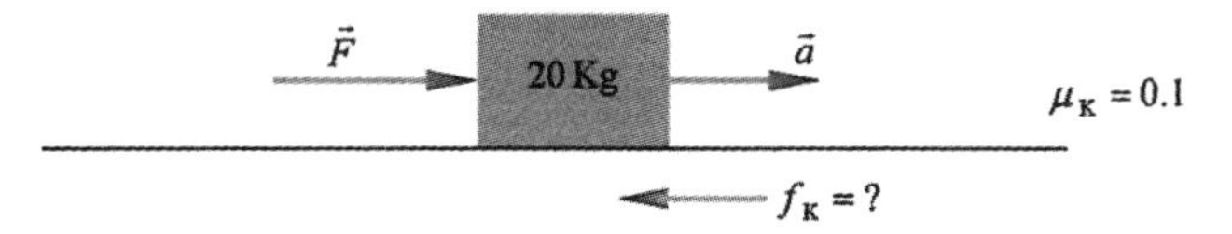

9. If the coefficient of static friction between the mass and the incline in Exercise 7 is 0.48, find the force of static friction between the mass and the incline.

10. In order for a mass to slide down an incline, the component of the weight of the mass directed down the plane must be larger than the force of static friction attempting to hold the mass in place. For the following system, determine with a calculation whether or not the mass will slide.

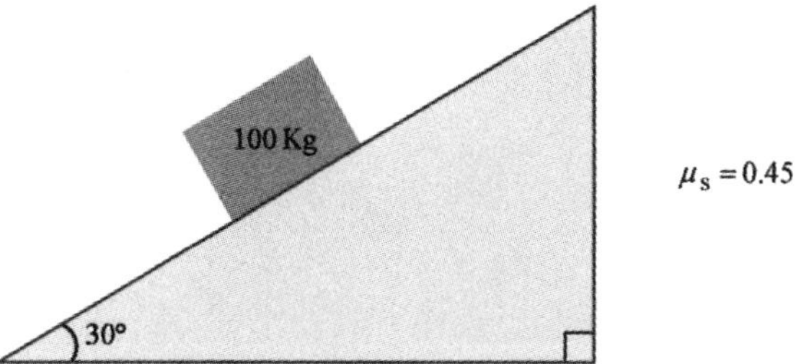

STRUCTURAL ANALYSIS AND DESIGN 6

Using a Rotated Coordinate System to Analyze a Design

This application examines a static situation using a rotated coordinate system.

> **Connection**
>
> The ability to analyze forces that are not exerted along the horizontal and vertical axes is necessary to understand trusses and many other architectural forms.
>
> Trusses are discussed in Structural Analysis and Design Application 11.

In each of the previous applications, the forces that were applied in the system were oriented along the conventional x- and y-axes. In this application we will extend our understanding of perpendicular forces by analyzing a system that has a more complex orientation of forces. As we will see, there are situations in which it is preferable to use a perpendicular coordinate system that is not strictly horizontal and vertical to solve the problem.

Solving Problems with a Standard Coordinate System

Suppose that we have the following static situation, depicted in the figure, in which we wish to determine both the tension in the cable and the magnitude of the normal force exerted on the mass by the inclined plane.

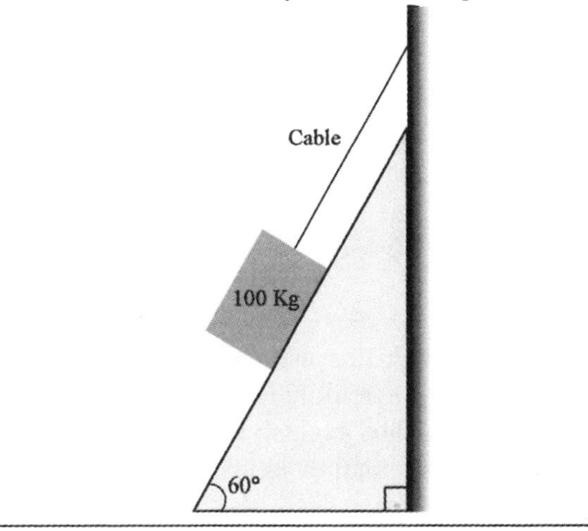

Figure 6.1

When we insert the three forces that act on the mass, we see that they are rather inconveniently oriented from the standpoint of easy vector analysis.

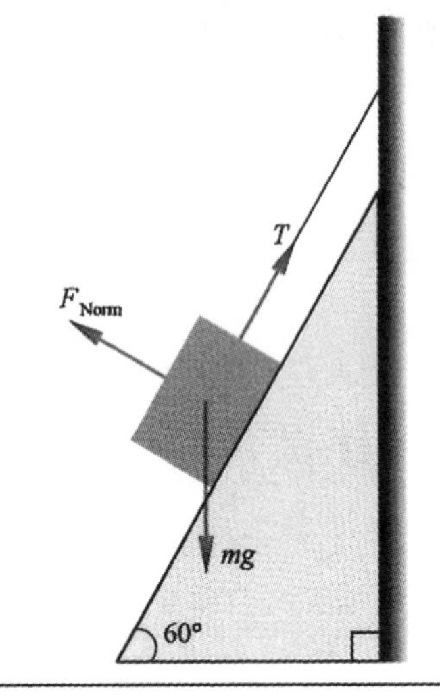

Figure 6.2

One approach to solving this system is to use a standard coordinate system with a horizontal *x*-axis and a vertical *y*-axis. Using the rules of vector analysis, we can then break each force vector into *x* and *y* components as shown in the figure:

> **Connection**
>
> For a refresher on breaking vectors into their perpendicular components, see Section 3.2 in *College Physics.*

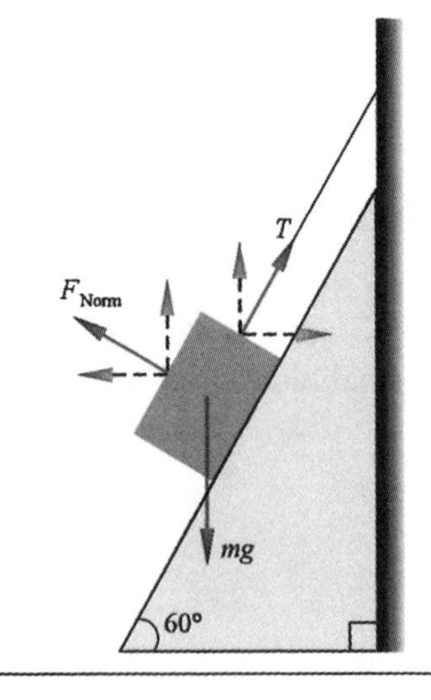

Figure 6.3

Although this method will yield a solution to the problem, the process is labor intensive and hides the intuitive, *physical* aspects of the problem. When approached exclusively by means of vector analysis, the solution becomes purely mathematical in nature rather than being the solution to a *physical* problem.

Solving Problems with a Perpendicular Coordinate System

An alternative approach is to construct a coordinate system that reflects the orientation of forces in our problem. Instead of using the traditional horizontal and vertical axes, we may use a perpendicular coordinate system that is rotated to match the forces in the problem.

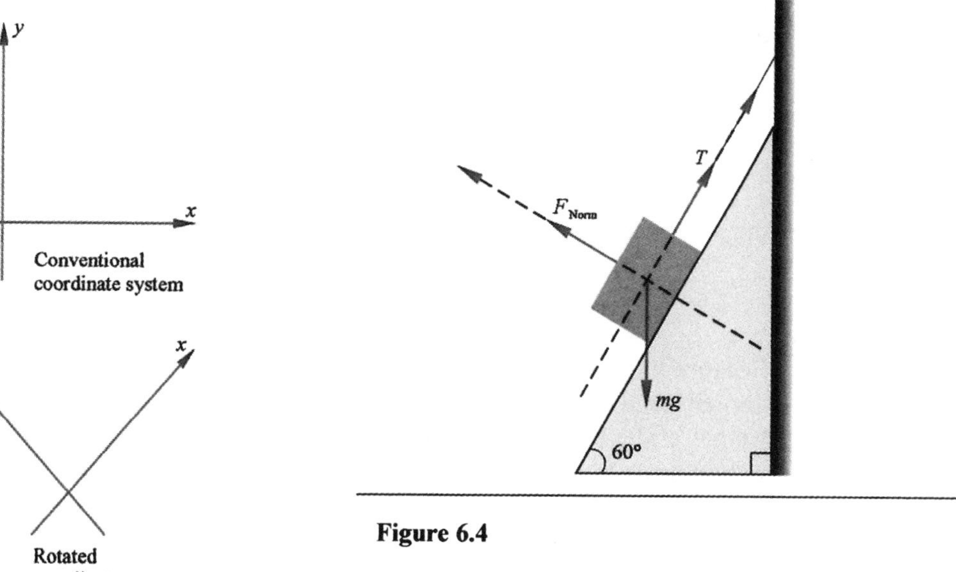

Figure 6.4

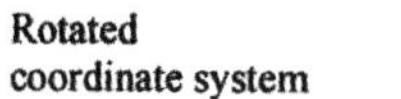

Figure 6.5

Using this coordinate system, only the weight vector must be resolved into components.

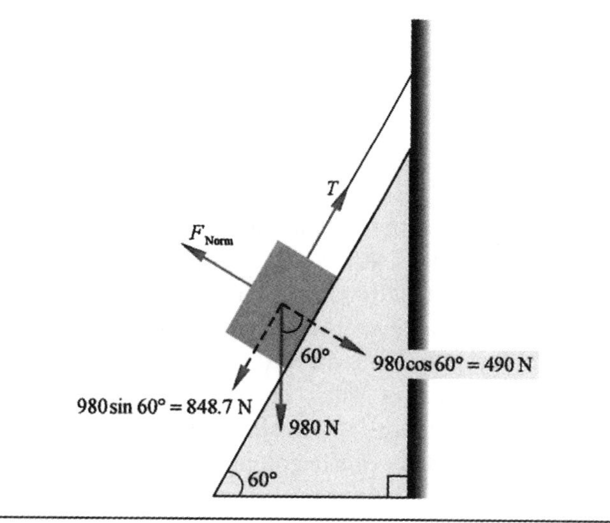

Figure 6.6

Drawing on our knowledge that the perpendicular axes are independent of one another allows us to use this rotated coordinate system to solve immediately for both the tension in the cable and the magnitude of the normal force exerted on the mass by the inclined plane.

1. Because the mass is not moving up or down the plane, the tension in the cable must be equal in magnitude to the component of the weight of the mass pointing down the plane. Since the component of the weight down the plane is

$$mg\sin\theta = (100\ \text{Kg})(9.8\ \text{m/s}^2)\sin 60^\circ$$
$$mg\sin\theta = 848.7\ \text{N}$$

the magnitude of the tension in the cable must also be

$$T = 848.7\ \text{N}$$

2. Because the mass is not falling into the plane or flying off the plane, the normal force exerted by the plane must be equal to the component of the weight of the mass oriented into the plane. Since the component of the weight directed into the plane is given by

$$mg\sin\theta = (100\ \text{Kg})(9.8\ \text{m/s}^2)\cos 60^\circ$$
$$mg\sin\theta = 490\ \text{N}$$

the magnitude of the normal force becomes

$$F_{\text{Norm}} = 490\ \text{N}$$

Connection

For a discussion of normal forces, see Section 4.4 in *College Physics.*

Conclusion

This example shows how using a rotated coordinate system allows us to find two unknowns in a very simple manner. The alternative approach of working with the conventional horizontal and vertical axes requires a more difficult and time-consuming mathematical exercise to solve for these same unknowns. Rotating the coordinate axes to align with the forces in the problem is another useful tool for solving static problems. Because of the various geometries involved in architectural structures, a designer must be able to analyze forces that are not necessarily exerted along the horizontal and vertical axes.

Exercises

1. For the following mass, find

   a) the component of the weight of the mass directed down the plane
   b) the tension in the cable

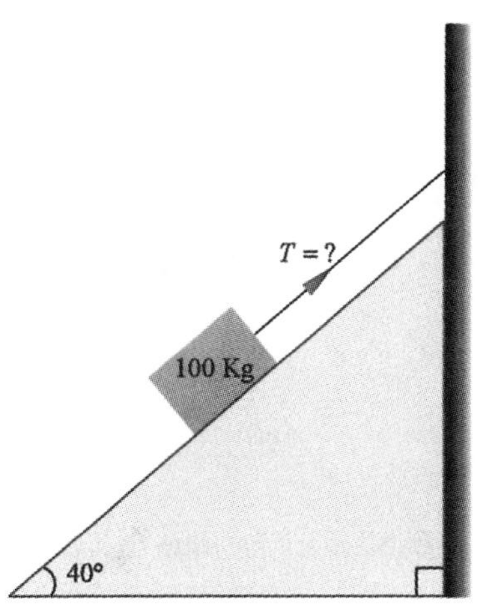

2. Calculate the tension in the cable and the normal force exerted by the inclined plane on the mass using a coordinate system that is rotated to match the forces in the problem.

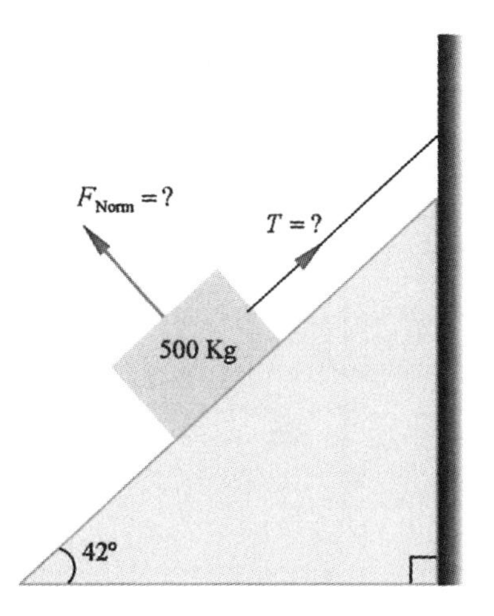

3. Find the normal force exerted on each mass and the tension in each cable. Assume that the incline is frictionless.

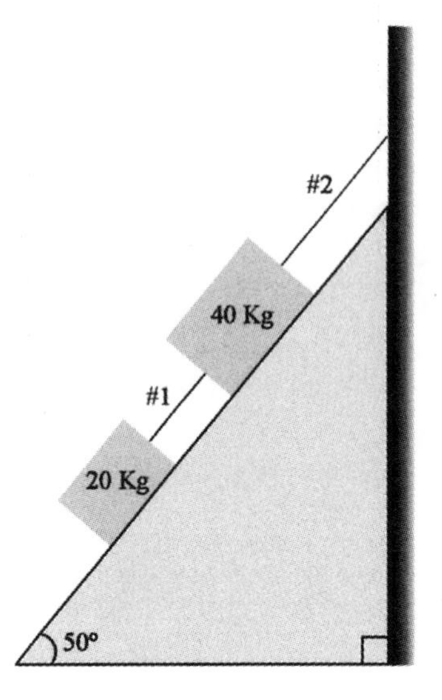

4. Find the tension in cable #2. Assume that the pulley and the incline are frictionless.

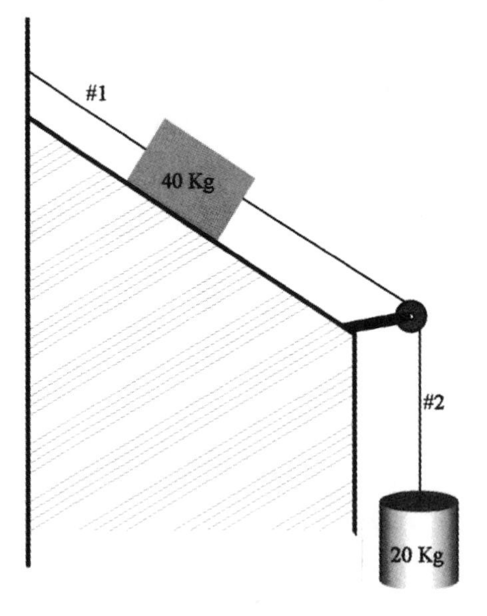

5. Find the tension in the cable.

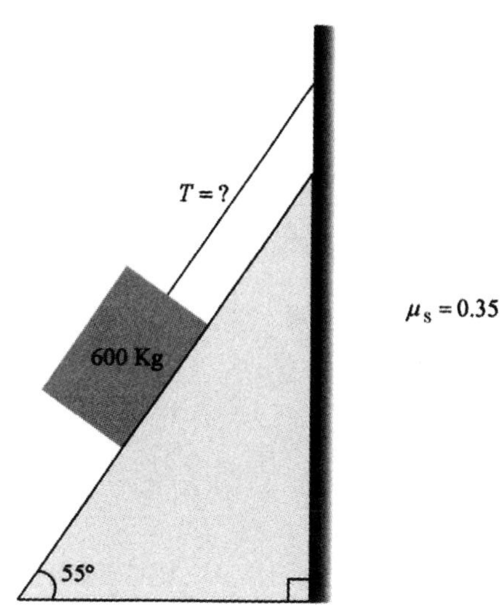

STRUCTURAL ANALYSIS AND DESIGN 7

Systems that are Static Along Both the X- and Y-Axes

This application provides an opportunity to study a system that is static along both the x- and y-axes. As we will see, the physical realization that the system is not moving along either axis will allow us to find two unknown quantities in the system simultaneously.

In previous applications, we saw that Newton's Second Law provides a powerful tool for analyzing fixed, non-moving situations. Let's take a moment to review what we learned.

> **Review**
>
> Newton's Second Law:
>
> $\sum \vec{F} = m\vec{a}$

Newton's Second Law relates the amount of force exerted on a mass to the acceleration of the mass. If an object is not moving, or is *static*, it must have an acceleration equal to zero. If the acceleration of an object is zero, the expression for Newton's Second Law reduces to

$$\sum \vec{F} = 0$$

We can take the force vectors in Newton's law and analyze each component individually.

If an object is not moving to the left or to the right, then the forces along the horizontal axis must be adding to zero. If there were a net force along the horizontal axis, the object would then move in the direction of this applied force. Similarly, if an object is not moving upward or downward, the forces applied along the vertical axis must add to zero.

In equation form,

$$\sum F_x = 0$$
$$\sum F_y = 0$$

Although these equations appear to be simple, they are sufficiently powerful to solve some extremely complex physical problems.

▸ The Hanging Traffic Light

Use the illustration in the figure to find the tension in each cable that supports the traffic light.

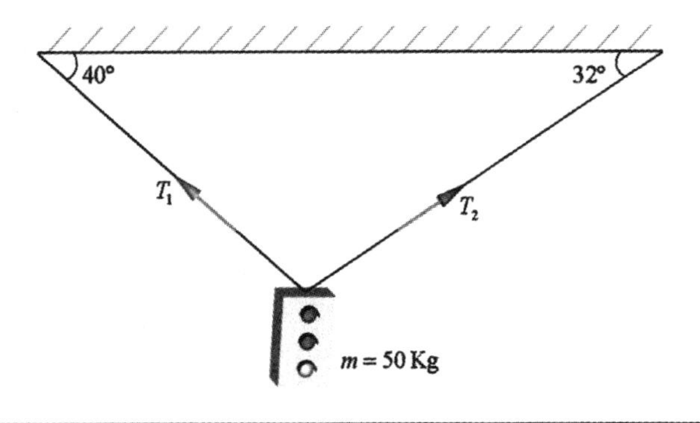

Figure 7.1

Connection

For a refresher on breaking vectors into components, see Section 3.2 of *College Physics.*

Solution The first step in solving for the tension in the cables is to separate each tension vector into its horizontal and vertical components and then to insert the weight vector for the traffic light as shown:

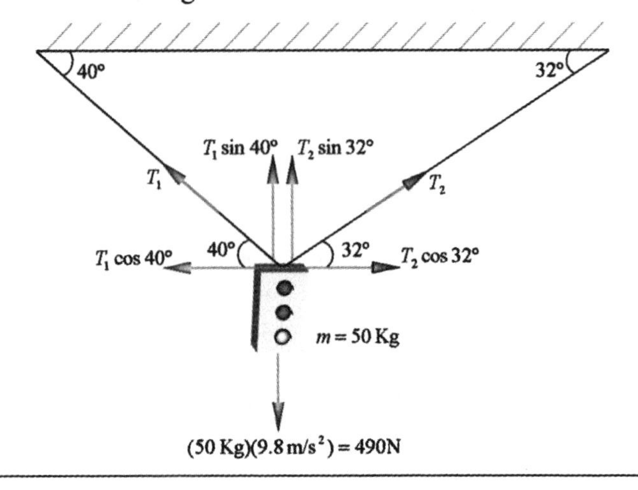

Figure 7.2

Because the traffic light does not move to the left or right, we know that the forces along the x-axis must total to zero. Thus,

$$\sum F_x = 0$$

becomes

$$T_2 \cos 32° - T_1 \cos 40° = 0$$

Similarly, because the streetlight does not move upward or downward,

$$\sum F_y = 0$$

becomes

$$T_1 \sin 40° + T_2 \sin 32° - 490 = 0$$

The two equations both have the same unknowns, T_1 and T_2, and can be solved simultaneously.

Begin with the horizontal equation,

$$\begin{aligned} &T_2 \cos 32° - T_1 \cos 40° = 0 \\ &T_2 \cos 32 = T_1 \cos 40° \\ &T_2 = T_1 \cdot \frac{\cos 40°}{\cos 32°} \\ &T_2 = 0.903 T_1 \qquad \textbf{(1)} \end{aligned}$$

Inserting this result into the vertical equation yields

$$\begin{aligned} &T_1 \sin 40° + T_2 \sin 32° - 490 = 0 \\ &T_1 \sin 40° + (0.903 T_1) \sin 32° - 490 = 0 \\ &0.643 T_1 + 0.478 T_1 = 490 \\ &1.121 T_1 = 490 \\ &T_1 = 437.11 \text{ N} \end{aligned}$$

Finally, reinserting the value for T_1 into Equation (1) yields the value for T_2:

$$T_2 = 0.903T_1$$
$$T_2 = 0.903(437.11)$$
$$T_2 = 394.71\,\text{N}$$

Conclusion

The preceding example is typical of static problems. The equations for the sum of forces along the horizontal and vertical axes are constructed and solved simultaneously. This technique is often used to solve a design problem in which we have two unknown quantities.

Exercises

1. Explain, in physical terms, why in static situations we are able to make the mathematical assumption that $\Sigma F = 0$.

2. Find T_1 and T_2.

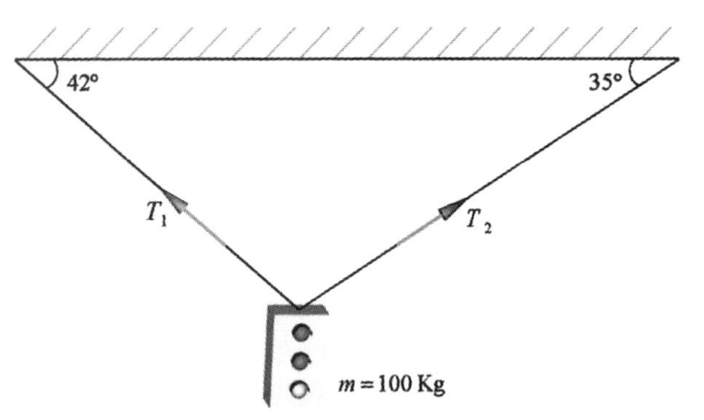

3. Find the force exerted on the sphere at point #1 and point #2. Hint: How is the weight distributed?

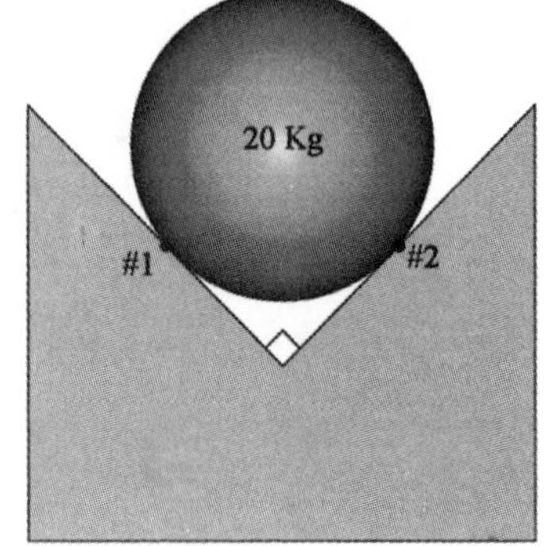

4. Find T_1 and θ_2.

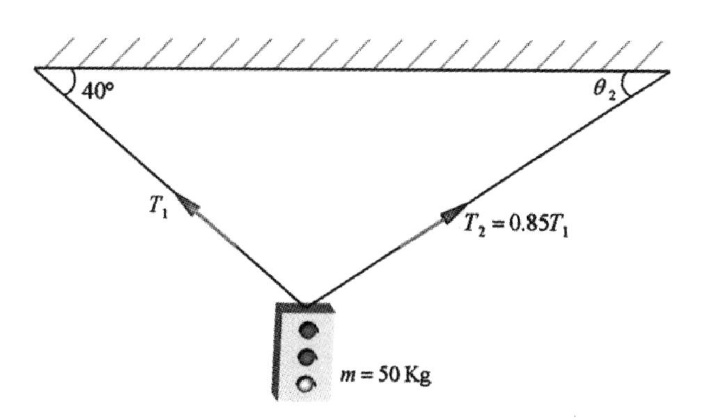

5. Find the tension in each cable.

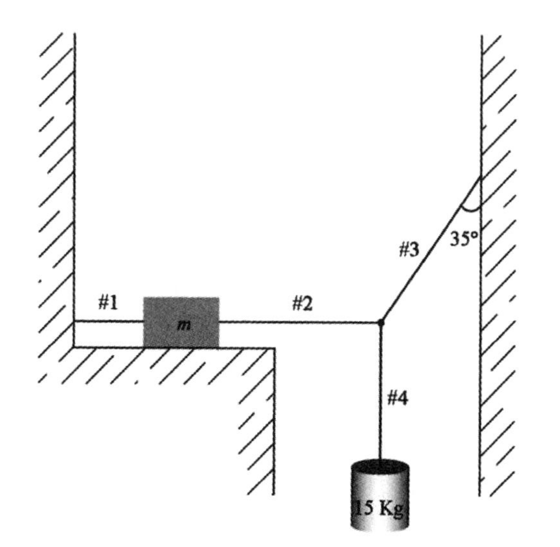

STRUCTURAL ANALYSIS AND DESIGN 8

Torque

In this application, we discuss the concepts of torque and rotational equilibrium.

> **Connection**
>
> The concepts of torque and rotational equilibrium are introduced in Section 8.2 of *College Physics.*

In Structural Analysis and Design Applications 3-7, we analyzed how forces cause objects to move in a linear direction and the conditions necessary to prevent objects from moving. In this application our focus is on the forces that cause objects to *rotate* rather than on those that cause objects to move in straight lines.

Torque

One way to visualize this concept is to think of a wrench that is fixed to a nut:

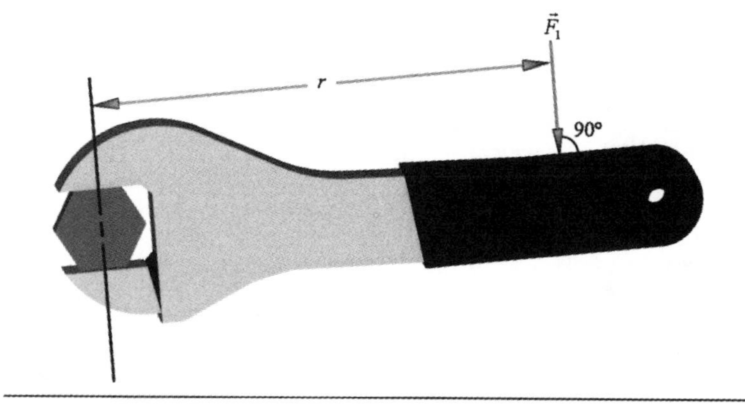

If a force, $\vec{F}_1$, is applied to the handle of the wrench, it will cause the handle to rotate. Experience tells us that it is easier to rotate the handle of the wrench when we apply the force closer to the end of the wrench. Similarly, we know that it is easier to open a door when we push at the handle rather than the hinge. The name for the distance from which the force is applied to the point of rotation (in our illustration, the nut) is the *lever arm*.

As we have seen, both the size of the force and the length of the lever arm affect the rotation of the handle. The product of the lever arm, r, and the force, $\vec{F}$, is called *torque* (τ).

$$\tau = rF \qquad \textbf{(1)}$$

However, we need to enhance Equation (1) to handle all possibilities. For example, if we apply a force, $\vec{F}_2$, to the handle, it may cause the handle of the wrench to move to the left, but it does not cause any rotation. Since $\vec{F}_2$ is parallel to the handle, it does not provide any torque to the handle of the wrench.

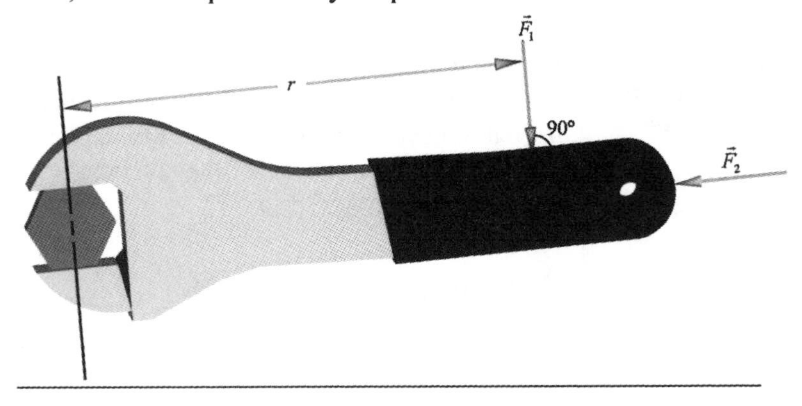

Figure 8.2

Connection

For a review on how to resolve a vector into components, see Section 3.2 of *College Physics*.

If we apply a force, $\vec{F}_3$, to the handle and resolve this force into its horizontal and vertical components, we see that $F_3\cos\theta$ behaves like $\vec{F}_2$ and $F_3\sin\theta$, the perpendicular component, behaves like $\vec{F}_1$.

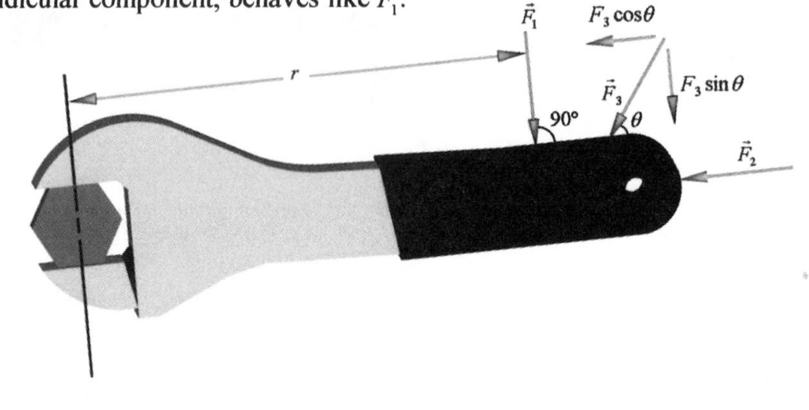

Figure 8.3

In other words, only the component of $\vec{F}_3$ that is perpendicular to the lever arm causes torque. As a result, we need to insert a $\sin\theta$ term into Equation (1).

$$\tau = Fr\sin\theta \qquad \textbf{(2)}$$

This additional term is important because it isolates the portion of the force that is at a right angle to the lever arm.

Before we begin to calculate the torque in our example, we must also take into consideration the fact that there are two possible directions for the torque on the handle of the wrench.

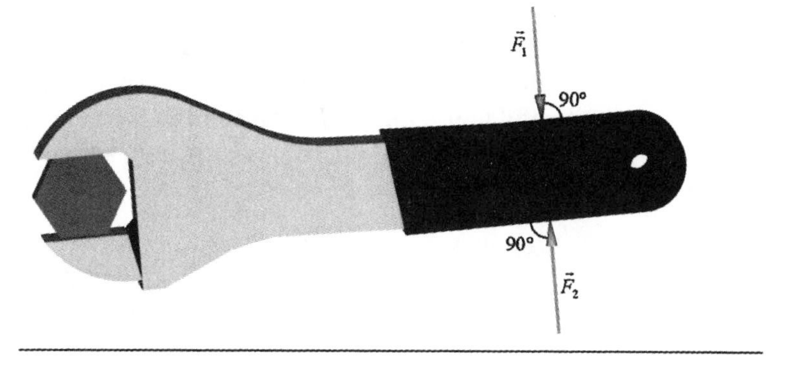

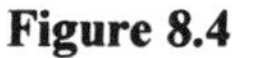

Figure 8.4

In Figure 8.4, we see that $\vec{F}_1$ will cause the handle of the wrench to rotate in a clockwise direction, while $\vec{F}_2$ will cause a counterclockwise rotation. We identify these different directions as having either positive or negative torques. As in the case of the forces, the assignment of the terms *positive* and *negative* depends on which direction we identify as being positive during the course of our analysis.

Example 1

For the strut illustrated in Figure 8.5, calculate the torque caused by the 10 N force, the torque caused by the 20 N force, and the total clockwise torque due to these two forces.

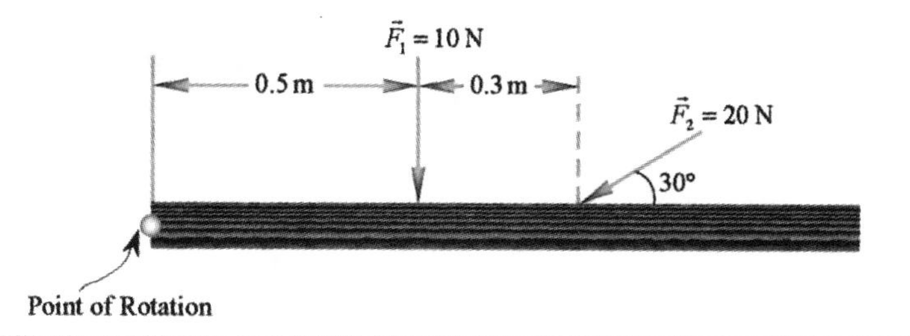

Figure 8.5

Finding the Torque Caused by the 10 N Force To find the torque caused by the 10 N force, we insert the force, the distance to the point of rotation, and the angle that the force makes with the lever arm into Equation (2).

$$\tau = Fr\sin\theta$$
$$\tau = (10\ \mathrm{N})(0.5\ \mathrm{m})\sin 90^\circ$$
$$\tau = 5\ \mathrm{N}\cdot\mathrm{m}$$

Because sin 90° = 1, we see that when the force is exerted at a right angle to the lever arm, the torque calculation reduces to a simple multiplication problem in which the force is multiplied by the length of the lever arm.

Finding the Torque caused by the 20 N Force We will again use Equation (2) to find the torque caused by the 20 N force. In this case, however, we must be careful to use the *total* distance from the point at which the 20 N force is applied measured back to the point of rotation, and not merely the distance between the 20 N force and the 10 N force.

$$\tau = Fr\sin\theta$$
$$\tau = (20\ \mathrm{N})(0.8\ \mathrm{m})\sin 30^\circ$$
$$\tau = 8\ \mathrm{N}\cdot\mathrm{m}$$

Finding the Total Clockwise Torque Finally, to find the total torque caused by both forces, we simply add the answers found in the first two steps.

$$\tau_{\mathrm{Total}} = 5\ \mathrm{N}\cdot\mathrm{m} + 8\ \mathrm{N}\cdot\mathrm{m}$$
$$\tau_{\mathrm{Total}} = 13\ \mathrm{N}\cdot\mathrm{m}$$

Example 2

Calculate the clockwise torque resulting from the weight of the uniform strut.

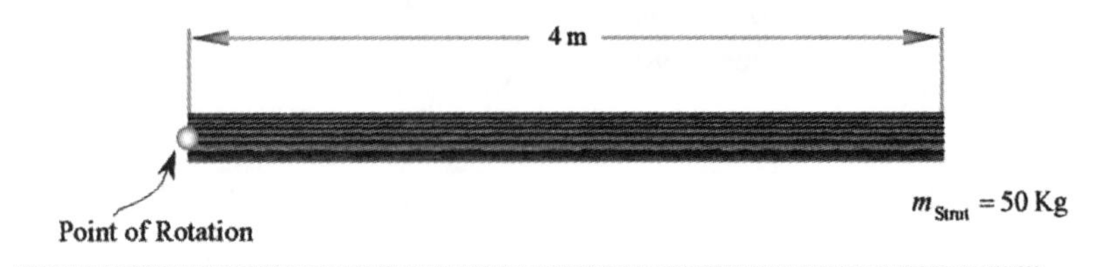

Figure 8.6

Solution Because the strut is uniform, we assume that the center of mass is located at the middle of the strut. We can also assume that the weight of the strut acts at the same point.

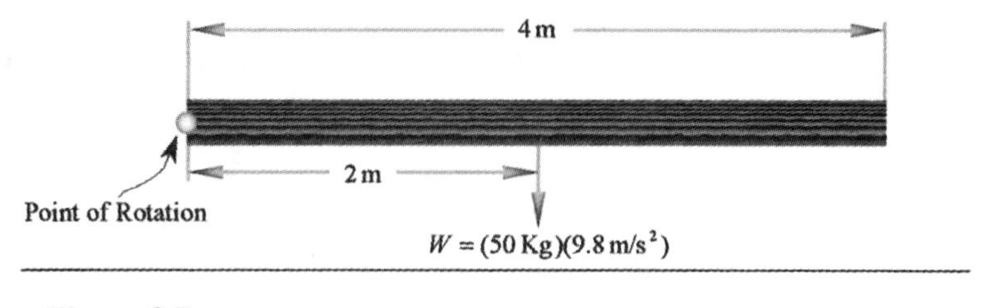

Figure 8.7

Based on our assumption, we calculate the length of the lever arm to be 2 m. Because the strut is horizontal, the weight vector makes a 90° angle with the lever arm. Using Equation (2), we insert the weight of the strut and the length of the lever arm to determine our answer.

$$\tau = Fr \sin \theta$$
$$\tau = (50\ \text{Kg})(9.8\ \text{m/s}^2)(2\ \text{m})\sin 90°$$
$$\tau = 980\ \text{N} \cdot \text{m}$$

After this short introduction to torque, let's now analyze a situation that contains equal amounts of both clockwise and counterclockwise torques.

Rotational Equilibrium

In Applications 3-7, we examined situations in which the sum of forces along a particular axis was zero. When the forces along an axis total to zero, we say that the system is in *translational equilibrium*. A similar term is used to denote a system in which no net torque is acting. When a system does not rotate, we say that it is in *rotational equilibrium*. For a system to be in rotational equilibrium the sum of the clockwise and counterclockwise torques must equal zero.

Example 3

For the strut introduced in Example 1 and illustrated in Figure 8.8, find the force $\vec{F}_3$ that is necessary to place the system in rotational equilibrium.

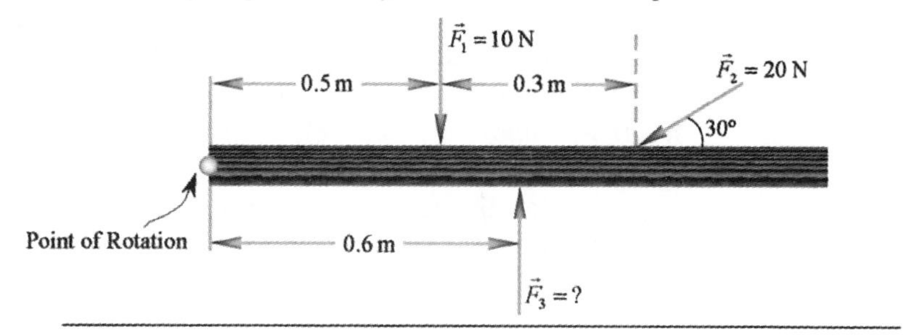

Figure 8.8

Solution We know that for the system to be in rotational equilibrium, the torque generated by $\vec{F}_3$ must be the same as the combined torque caused by $\vec{F}_1$ and $\vec{F}_2$, 13 Nm.

Using Equation (2) we insert the given information

$$\tau = Fr\sin\theta$$

$$13\,\text{N}\cdot\text{m} = F_3(0.6\,\text{m})\sin 90^\circ$$

$$\frac{13\,\text{N}\cdot\text{m}}{0.6\,\text{m}} = F_3$$

$$F_3 = 21.67\,\text{N}$$

and find that a force of 21.67 N is required to place the system in rotational equilibrium.

Choosing the Point of Rotation

In all of the previous examples, the point about which we measured the torques (the fulcrum) was already chosen. Defining the point initially allowed us to introduce the concept of torque more simply. In actuality, if an object is in rotational equilibrium we have the freedom to choose any point on the object as the point of rotation.

The uniform beam in Figure 8.9 is in rotational equilibrium. Because it is not rotating, the sum of the torques about any point on the beam must be zero. For example, the clockwise and counterclockwise torques about point A must be equal to one another. Otherwise, the beam would rotate around point A. We can make similar arguments at points B and C.

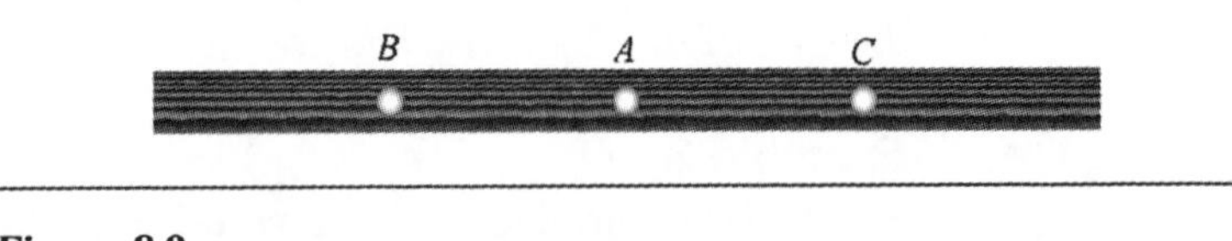

Figure 8.9

We can use the fact that we have freedom to choose the point about which to measure the torques in a static situation to our advantage. In Figure 8.10, notice that if we choose the point of rotation to be point A, the center of the beam, the weight of the beam will not enter the torque calculation. Because there is no perpendicular distance from where the force is being applied measured from the point of rotation, there is no lever arm for this force. Therefore, the weight of the beam will not enter the sum-of-torques equation when point A is chosen as the point of rotation.

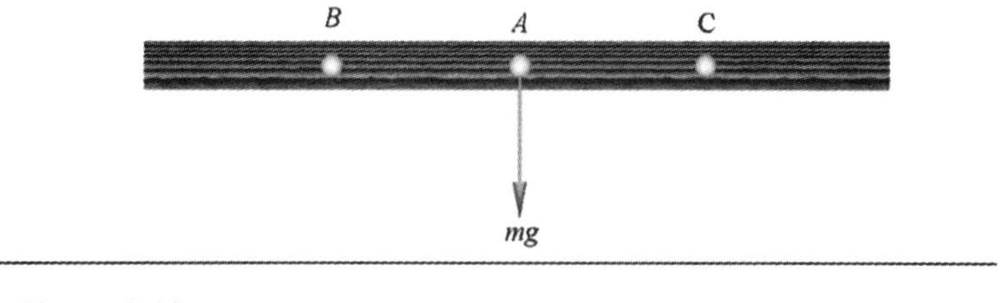

Figure 8.10

However, if we instead choose point B to be the point of rotation, the weight of the beam will now enter the torque calculation since the force is being exerted a perpendicular distance from the point of rotation.

Conclusion

An understanding of torque is necessary for all students of design. In order for a structure to be stable, the torques caused by all of the beams, struts, and other building materials must be clearly analyzed and understood. As we saw in this application, when executing this analysis, it is advantageous to choose the location of the point of rotation to be at the same point as *at least* one of the forces being exerted in the problem. This choice eliminates one force from the sum-of-torques equation and greatly simplifies the algebra.

Exercises

1. Calculate the torque caused by the force shown in the figure.

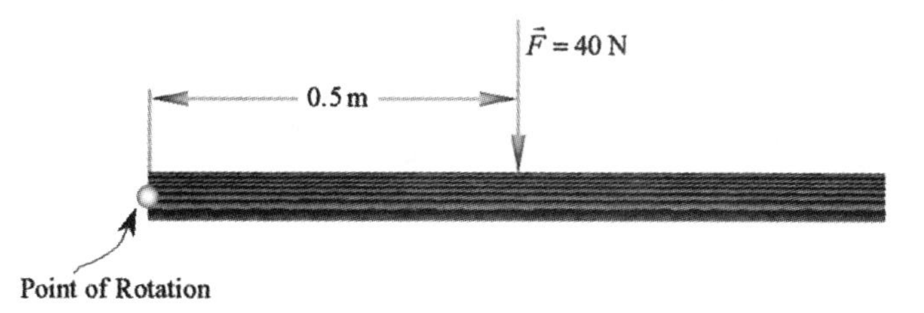

2. Calculate the total clockwise torque caused by $\vec{F}_1$ and $\vec{F}_2$.

$\vec{F}_1 = 35\,\text{N}$
$\vec{F}_2 = 50\,\text{N}$
0.2 m
40°
Point of Rotation
0.3 m

3. Find the net torque on the lever arm.

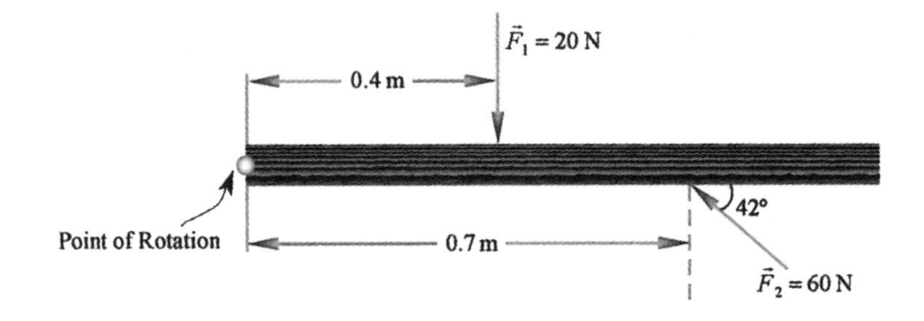

4. Find the magnitude of $\vec{F}_3$ required to put the lever arm into rotational equilibrium.

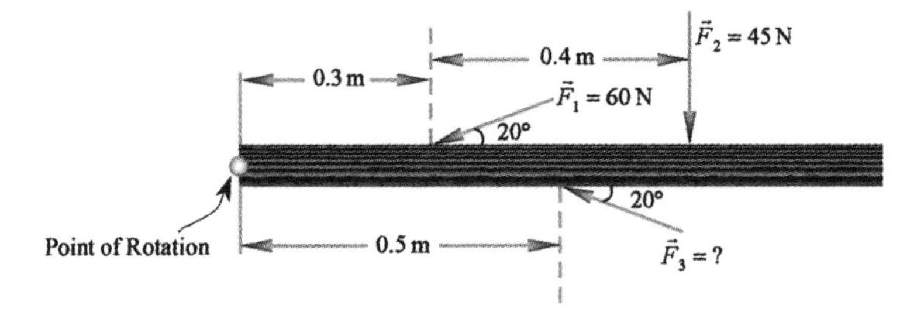

5. Find the torque on the uniform strut resulting from the weight of the strut.

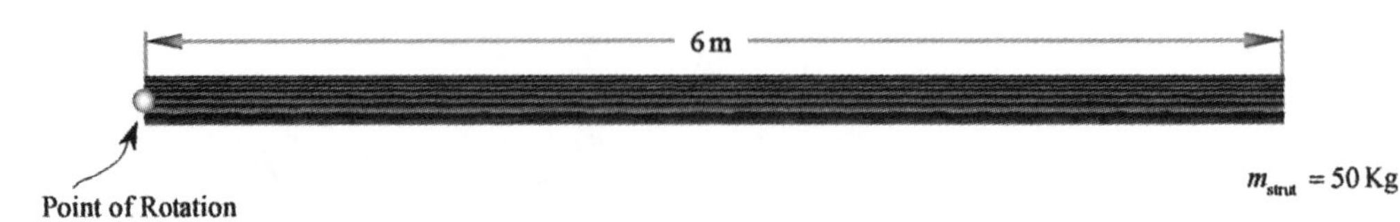

6. Find the magnitude of $\vec{F}$ required to place the 50 Kg strut into rotational equilibrium.

$m_{strut} = 50$ Kg

2 m

1.75 m

60°

$\vec{F}$

Point of Rotation

7. Using the fact that the strut is in rotational equilibrium, find the tension in the cable. Assume that the strut is uniform.

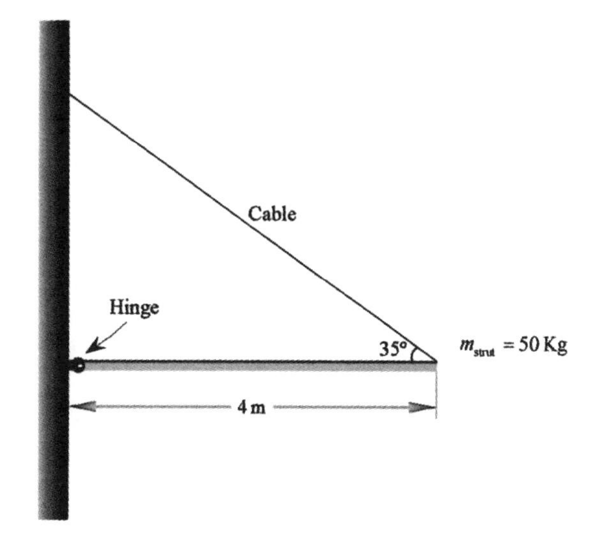

STRUCTURAL ANALYSIS AND DESIGN 9

Rotational and Translational Equilibrium: A System with Three Unknowns

This application brings together the concepts of translational and rotational equilibrium discussed in Structural Analysis and Design Applications 3-8 to show how to solve a static situation with three unknowns.

Connection

It may be helpful to work through Structural Analysis and Design Applications 3-8 before attempting this application.

In Applications 3-7, we examined situations in translational equilibrium in which the forces added to zero, and in Application 8 we examined examples of rotational equilibrium in which the torques added to zero. Now let's look at a more complex problem that uses both of these concepts to solve a static situation that has three unknown quantities. This type of analysis is common in the design process since there are usually several unknown physical quantities that must be found and interpreted.

Solving a System with Three Unknowns

From algebra, we know that solving a system with three unknowns requires us to have a system of three equations that relate these unknown quantities. In this case, the three equations that describe our static situation tell us that the system is not moving horizontally:

$$\sum F_x = 0$$

nor is it moving vertically:

$$\sum F_y = 0$$

nor is it rotating:

$$\sum \tau = 0$$

Let's take these three equations and use them to solve an actual physical system that has three unknowns.

The Hanging Strut: A System in Translational and Rotational Equilibrium

Suppose that we have the following situation in which a 50 Kg uniform strut is attached to a hinge with an additional mass of 20 Kg hanging from it. Using the information from Applications 3-8, let's find the tension in the support cable and the forces exerted on the strut by the supporting hinge.

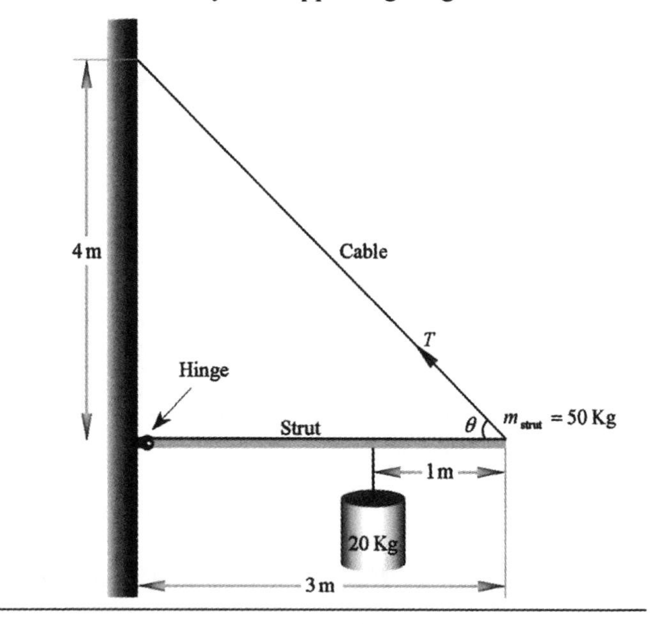

Figure 9.1

The Angle that the Support Cable Makes with the Strut

From the illustration, we see that the tension in the cable is supporting the right end of the strut. Although we do not know a value for the tension in the cable, we do know how to break force vectors into their perpendicular components. Since the cable, strut, and post form a right triangle, the angle between the cable and the strut can be found from trigonometry:

$$\theta = \tan^{-1}\left(\frac{4}{3}\right)$$

$$\theta = 53.13°$$

The Components of the Tension in the Support Cable

This angle can now be used to resolve the tension into its perpendicular components.

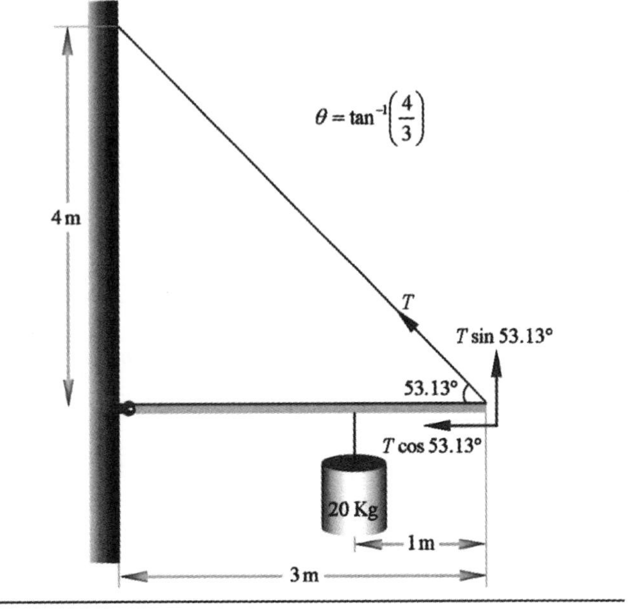

Figure 9.2

The Horizontal and Vertical Forces on the Strut Due to the Hinge

Because the strut is not moving to the left, we know that $T\cos 53.13°$ cannot be the only force acting along the horizontal axis. An opposing force provided by the hinge must be present to keep the strut static along this axis:

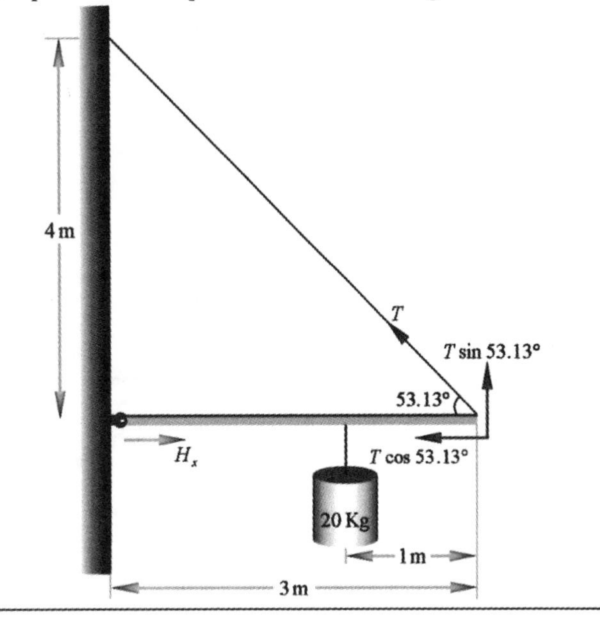

Figure 9.3

Additionally, we know that if the hinge were not present in the system, the left end of the strut would fall down. To keep the left end of the strut from dropping downward, the hinge must exert an upward force on this end of the strut:

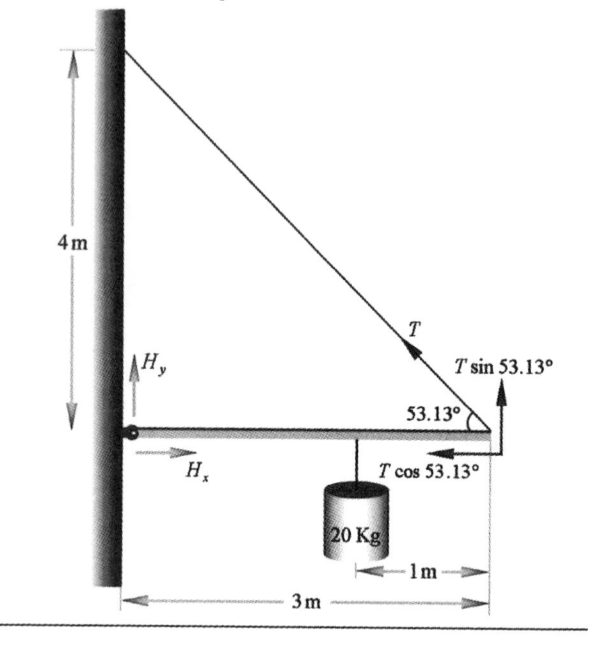

Figure 9.4

Taking into account the weight of the 20 Kg mass and the strut itself, we now have all of the forces acting on the strut.

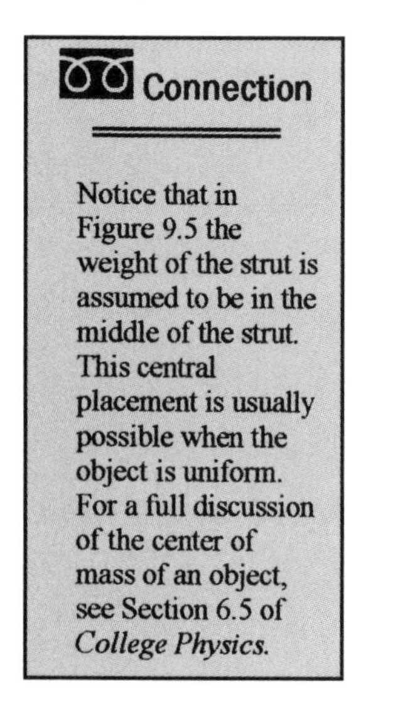

Connection

Notice that in Figure 9.5 the weight of the strut is assumed to be in the middle of the strut. This central placement is usually possible when the object is uniform. For a full discussion of the center of mass of an object, see Section 6.5 of *College Physics*.

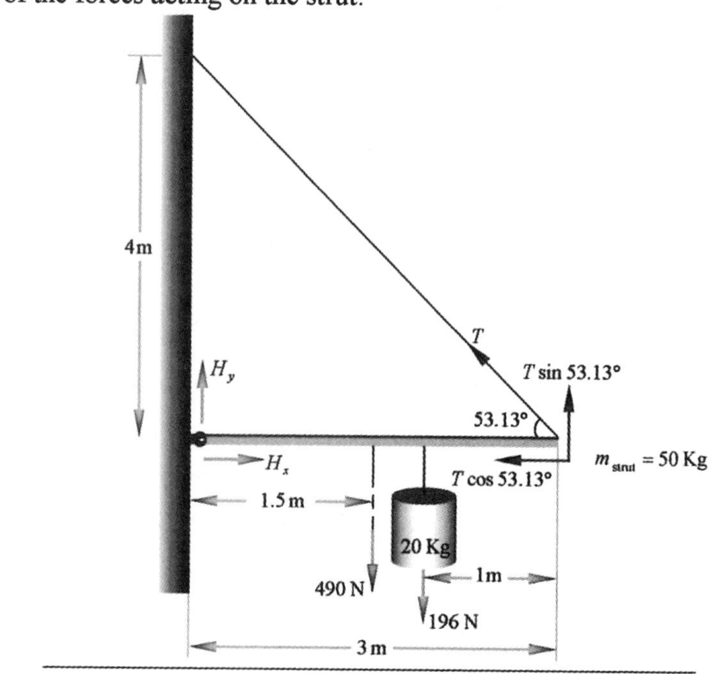

Figure 9.5

Constructing the Sum-of-Forces and Sum-of-Torques Equations

To now solve for the three unknown quantities, H_x, H_y, and T, we must find three simultaneous equations that relate these unknowns.

By choosing positive directions for x, y, and the torque, we can set up the two sum-of-forces equations and the sum-of-torques equation that will allow us to solve for the three unknowns. The two sum-of-forces equations take the form

$$\sum F_x = 0$$

$$H_x - T\cos 53.13° = 0$$

$$\sum F_y = 0$$

$$H_y + T\sin 53.13° - 490\,\text{N} - 196\,\text{N} = 0$$

> **Connection**
>
> For a discussion on the choice of the point of rotation, see Structural Analysis and Design Application 8.

Before we can construct a sum-of-torques equation, we must designate a point of rotation. Because the strut is not rotating, we have complete freedom as to which point along the strut we choose to identify as the point of rotation, or fulcrum. In Application 8, we learned that we can eliminate one of the forces in the torque equation by choosing the fulcrum to be a point along the strut at which one of the forces in the problem is being exerted. Consequently, in this example, if we choose the hinge as the point of rotation, neither H_x, nor H_y will enter the torque equation, thus greatly simplifying the algebra.

Choosing the hinge as the point of rotation, the torque equation becomes

$$\sum \tau = 0$$

$$(T\sin 53.13°)(3\,\text{m}) - (490\,\text{N})(1.5\,\text{m}) - (196\,\text{N})(2\,\text{m}) = 0$$

By strategically choosing the point of rotation, we have reduced the sum-of-torques equation to an equation with only one unknown that we can now solve directly for the tension in the supporting cable. Notice that the torques caused by the weight of the strut and the weight of the 20 Kg mass do not explicitly contain the sine of an angle in their expressions because the weight vector for each makes a 90° angle with the strut.

Solving the System of Equations

Solving the torque equation for the tension in the cable gives us

$$(T \sin 53.13°)(3\,\text{m}) - (490\,\text{N})(1.5\,\text{N}) - (196\,\text{N})(2\,\text{m}) = 0$$
$$2.4T - 735 - 392 = 0$$
$$2.4T - 1127 = 0$$
$$2.4T = 1127$$
$$T = 469.6\,\text{N}$$

The tension can now be inserted into the horizontal and vertical force equations to yield H_x:

$$H_x - T\cos 53.13° = 0$$
$$H_x = T\cos 53.13°$$
$$H_x = (469.6\,\text{N})\cos 53.13°$$
$$H_x = (469.6\,\text{N})(0.6)$$
$$H_x = 281.8\,\text{N}$$

and H_y:

$$H_y + T\sin 53.13° - 490\,\text{N} - 196\,\text{N} = 0$$
$$H_y = 490\,\text{N} + 196\,\text{N} - T\sin 53.13°$$
$$H_y = 686\,\text{N} - T\sin 53.13°$$
$$H_y = 686\,\text{N} - (469.6\,\text{N})\sin 53.13°$$
$$H_y = 686\,\text{N} - (469.6\,\text{N})(0.8)$$
$$H_y = 686\,\text{N} - 375.7\,\text{N}$$
$$H_y = 310.3\,\text{N}$$

▸ Conclusion

In this section we have learned how the concepts of translational and rotational equilibrium may be used to solve a static problem with three unknowns. By making a strategic choice in selecting our point of rotation, we have seen how the problem can be simplified. However, it is important to note that even if we had chosen a different point along the strut about which to measure the torques, we could still have solved the problem. Selecting a different point of rotation would simply have made the algebra slightly more difficult.

The ability to execute this type of analysis is a valuable tool for any designer or structural engineer. In real world designs, there are usually many different unknown quantities that the designer must take into account during the design process.

Exercises

1. Find the tension in the cable and the horizontal and vertical components of the force on the strut resulting from the hinge.

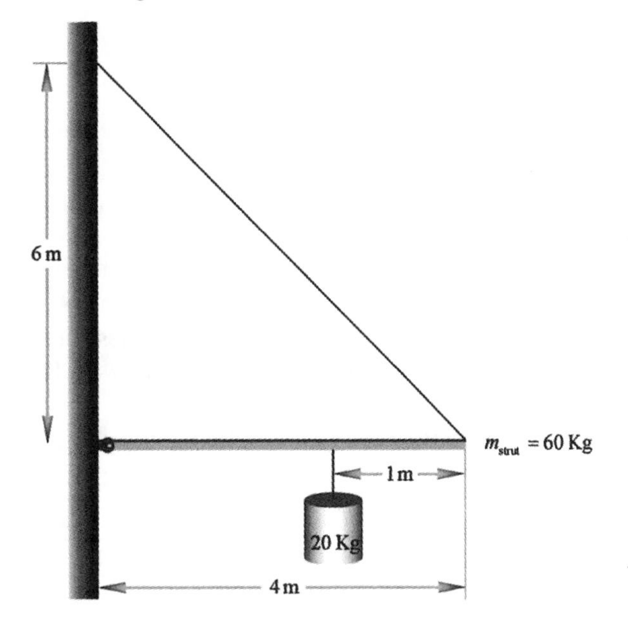

2. Knowing that the system is in rotational and translational equilibrium, find the center of mass of the non-uniform strut measured from the left side of the strut.

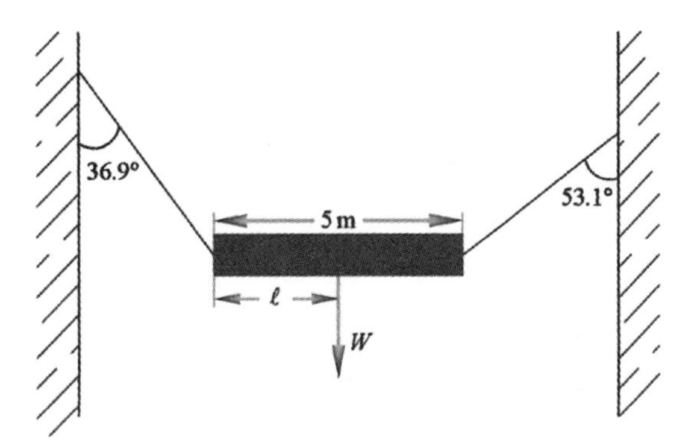

3. In the following system, the maximum tension that the cable can withstand is 3000 N.

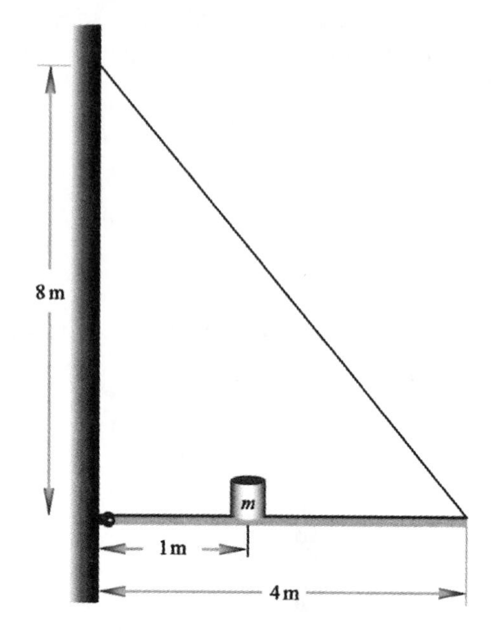

a) What is the maximum amount of mass that can be placed 1 m from the hinge so that the cable will not snap? Assume that the mass of the strut is negligible.
b) With the maximum amount of mass found in a), find the horizontal and vertical components of the force exerted by the hinge on the strut.

4. The system in the figure is in static equilibrium.

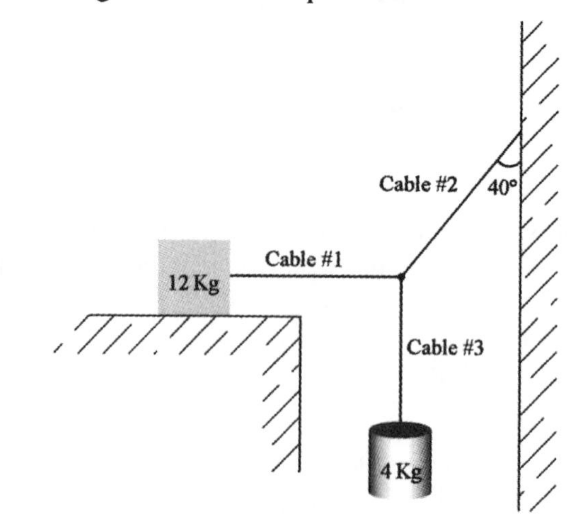

Find:

a) the tension in each cable
b) the force of static friction between the 12 Kg mass and the surface
c) the coefficient of static friction between the 12 Kg mass and the surface

STRUCTURAL ANALYSIS AND DESIGN 10

The Effects of Thermal Expansions in Static Situations

In this application, we discuss the effect of thermal expansions on materials and apply it to a static situation similar to that discussed in Structural Analysis and Design Application 9.

Connection

See *College Physics* Section 5.3 for a discussion of kinetic energy and Section 10.4 to learn about thermal expansion.

FYI

An example of a structure that must allow for thermal expansion and contraction is a bridge made from concrete and steel.

As discussed in Section 10.4 of *College Physics*, the temperature of a material, like structural steel, is proportional to the kinetic energy of its molecules. Because each of the molecules in a structural material vibrates farther from its central point as the temperature increases, the material expands. These expansions are called *thermal expansions*.

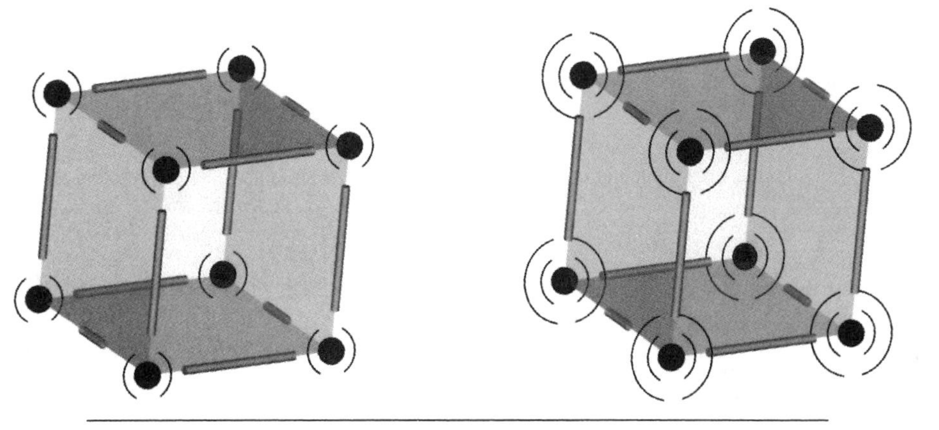

Figure 10.1

Linear Expansions

With some three-dimensional objects, one dimension is much larger than the other two. In such cases, we can reasonably approximate the object as having only one dimension. An example of such an object is a long piece of fishing line. Although in reality the line is a three-dimensional cylinder, it is reasonable to treat the line as a one-dimensional object possessing only length. In these cases, we can consider a simpler subset of thermal expansions called linear thermal expansions.

Linear thermal expansions involve an increase in the length of an object resulting from an increase in temperature. As the molecules in the object vibrate with more kinetic energy, the object gets longer.

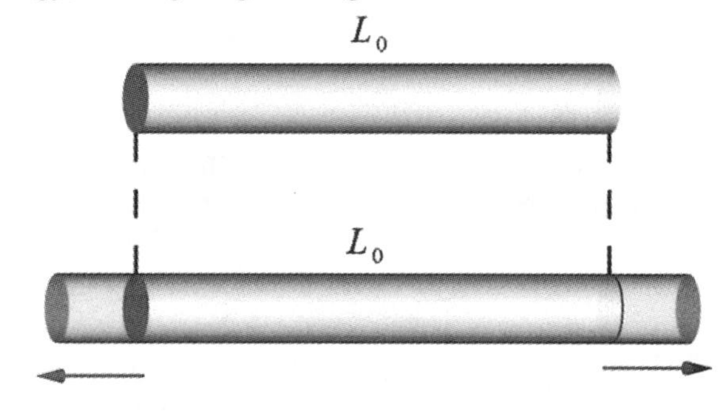

Figure 10.2

The amount of change in the length of the object is found using the equation

$$\Delta L = L_0 \alpha \Delta T \qquad \textbf{(1)}$$

where

- ΔL is the change in the length of the object
- L_0 is the original length of the object before the temperature change
- α is the coefficient of linear expansion of the material comprising the object
- ΔT is the change in the temperature of the object

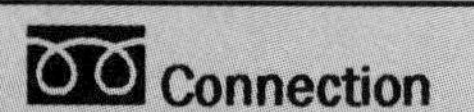

You can find a list of some coefficients of linear expansion in Table 10.1 of *College Physics.*

Notice that because the original length, L_0, and the change in the length, ΔL, are measured in the same units, the units on the coefficient of linear expansion and the change in the temperature cancel one another. Thus, if the change in the temperature is recorded in °C, the units on the coefficient of linear expansion must be recorded in 1/°C.

Equation (1) is used only to find the *change* in the length of the object. To find the new length of the object after the change in temperature, we must add the change to the original length if the material has increased in temperature ($\Delta T > 0$)

$$L_{\text{New}} = L_0 + \Delta L$$

or subtract the change if the material has decreased in temperature ($\Delta T < 0$)

$$L_{\text{New}} = L_0 - \Delta L$$

Example 1

Find the new length of a 2 m aluminum rod if the rod is raised from 10°C to 35°C.

Solution First, we find the change in the length using Equation (1) and the coefficient of linear expansion for aluminum, 24 x 10^{-6} 1/C°, found in Table 10.1 of *College Physics*.

$$\Delta L = L_0 \alpha \Delta T$$

$$\Delta L = (2\,\text{m})\left(24 \times 10^{-6} \frac{1}{°\text{C}}\right)(35°\text{C} - 10°\text{C})$$

$$\Delta L = 1.2 \times 10^{-3}\ \text{m}$$

Since the temperature of the rod increased, we find the new length by adding the change to the original length. Thus,

$$L_{\text{New}} = 2\,\text{m} + 0.0012\,\text{m}$$

$$L_{\text{New}} = 2.0012\,\text{m}$$

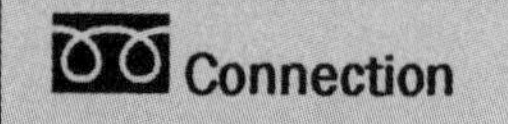

A sample of coefficients of volumetric expansion are found in Table 10.1 of *College Physics.*

Volumetric Expansions

Although one-dimensional approximations are sometimes useful, in most real world situations we must deal with the full three-dimensional expansion of an object. The equation used to find the change in the volume of an object is similar to the linear expansion equation and is given by

$$\Delta V = V_0 \beta \Delta T \qquad \textbf{(2)}$$

where:

- ΔV is the change in the volume of the object
- V_0 is the original volume of the object
- β is the coefficient of volumetric expansion of the material
- ΔT is the change in the temperature of the object

As with the coefficient of linear expansion, the units on the coefficient of volumetric expansion are 1/°C.

Example 2

A lead sphere has a radius of 0.25 m when at a temperature of 5°C. Find the new volume of the sphere if it is raised to 50°C. The coefficient of volumetric expansion for lead is 87 x 10^{-6} 1/°C.

Solution First, we find the original volume of the sphere using its given radius:

$$V_{\text{Sphere}} = \frac{4}{3}\pi r^3$$

$$V_{\text{Sphere}} = \frac{4}{3}\pi (0.25\ \text{m})^3$$

$$V_{\text{Sphere}} = 65.4 \times 10^{-3}\ \text{m}^3$$

Having found the original volume, we insert it along with the temperature change and the coefficient of volumetric expansion for lead into Equation (2)

$$\Delta V = V_0 \beta \Delta T$$

$$\Delta V = \left(65.4 \times 10^{-3}\ \text{m}^3\right)\left(87 \times 10^{-6} \frac{1}{°\text{C}}\right)\left(50°\text{C} - 5°\text{C}\right)$$

$$\Delta V = 2.56 \times 10^{-4}\ \text{m}^3$$

Adding the change in the volume to the original volume yields the result

$$V_{\text{New}} = V_0 + \Delta V$$
$$V_{\text{New}} = 65.4\times10^{-3}\ \text{m}^3 + 2.56\times10^{-4}\ \text{m}^3$$
$$V_{\text{New}} = 65.7\times10^{-3}\ \text{m}^3$$

▸ The Relationship Between Linear and Volumetric Expansion Coefficients

For most materials, the coefficient of linear expansion and the coefficient of volumetric expansion are related by

$$\beta = 3\alpha$$

This relationship makes sense when we think of a volumetric expansion as being three linear expansions happening simultaneously. For example, in Example 1 the coefficient of linear expansion for aluminum was given as 24 x 10^{-6} 1/°C. Thus, the coefficient of volumetric expansion for aluminum must be

$$\beta = 3\alpha$$
$$\beta = 3\left(24\times10^{-6}\frac{1}{°\text{C}}\right)$$
$$\beta = 72\times10^{-6}\frac{1}{°\text{C}}$$

With this brief background on the process of thermal expansion, we can now use our knowledge in a statics application similar to the one we examined in Application 9, *Rotational and Translational Equilibrium: A System with Three Unknowns.*

▸ The Effects of a Thermal Expansion on a Static Situation

Connection

It may be useful to review Structural Analysis and Design Application 9 before beginning this example.

Example 3

Consider a situation in which a 50 Kg uniform aluminum strut is attached to a hinge. The strut has a length of 3 m when it is at a temperature of 5°, as illustrated in Figure 10.3.

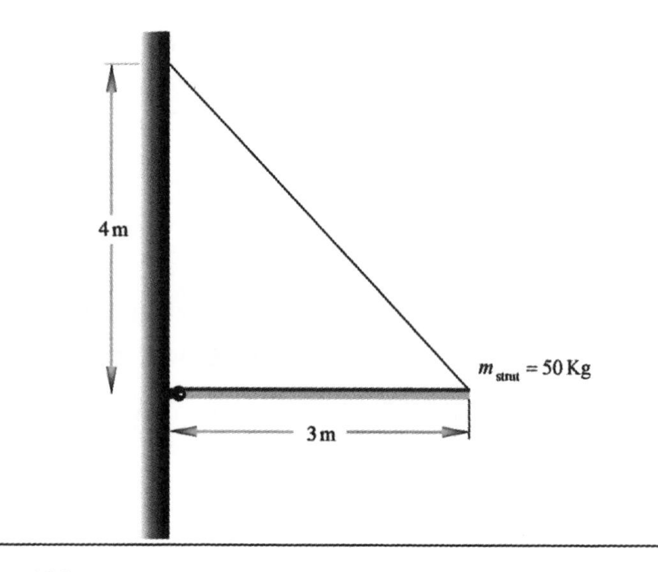

Figure 10.3

Our purpose in this example is to contrast a few of the characteristics in the figure, such as the center of mass of the strut and the tension in the cable, when the strut is at its starting temperature of 5°C and when it is raised to a temperature of 50°C.

For the sake of simplicity in our demonstration of the effects of thermal expansion, we will assume, somewhat unrealistically, that only the strut increases in temperature and not the hinge and cable as well. Further, to illustrate the small difference generated by thermal expansion, we will momentarily suspend the rules of significant figures.

Solution

Step1: The length of the strut at 50°C

To find the new length of the strut at 50°C we use Equation (1).

$$\Delta L = L_0 \alpha \Delta T$$

$$\Delta L = (3\text{ m})\left(24\times10^{-6}\,\frac{1}{°\text{C}}\right)(45°\text{ C})$$

$$\Delta L = 3.24\times10^{-3}\text{ m}$$

$$L_{\text{New}} = 3\text{ m} + 0.00324\text{ m}$$

$$L_{\text{New}} = 3.00324\text{ m}$$

Step 2: The angle of the cable at the two temperatures

At 5°C

The angle between the cable and the strut can be found using the length of the strut and the distance between the cable support and the hinge. To enable us to contrast the results at 5°C with those that at 50°C, we will show the angle measurement to several decimal places.

$$\theta = \tan^{-1}\left(\frac{4}{3}\right)$$

$$\theta = 53.130102°$$

At 50°C

At the higher temperature, we must use the expanded length of the strut in order to find the angle of the cable.

$$\theta = \tan^{-1}\left(\frac{4}{3.00324}\right)$$

$$\theta = 53.100412°$$

Review

The forces resulting from the hinge were identified as H_x and H_y in Application 9.

Because this angle is used in conjunction with the cable tension to find both the horizontal and vertical components of the force on the strut resulting from the hinge, these two solutions are also temperature-dependent.

Step 3: The center of mass of the strut at the two temperatures

If we assume that the strut is uniform, the center of mass can be found simply by finding the middle of the strut.

At 5°C

The center of mass is:

$$\text{center of mass} = \frac{3\text{ m}}{2} = 1.5\text{ m}$$

At 50° C

At this temperature, we must use the new length of the strut to find the new center of mass.

$$\text{center of mass} = \frac{3.00324\text{ m}}{2} = 1.50162\text{ m}$$

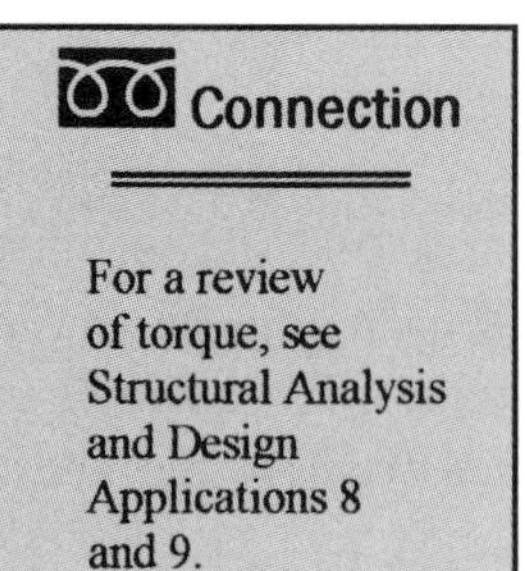

Connection

For a review of torque, see Structural Analysis and Design Applications 8 and 9.

Step 4: The tension in the cable at the two temperatures

The tension in the cable is found most easily by analyzing the torques on the strut with the hinge chosen as the point of rotation.

At 5°C

The vertical component of the tension provides the counterclockwise torque, while the weight of the strut provides the clockwise torque.

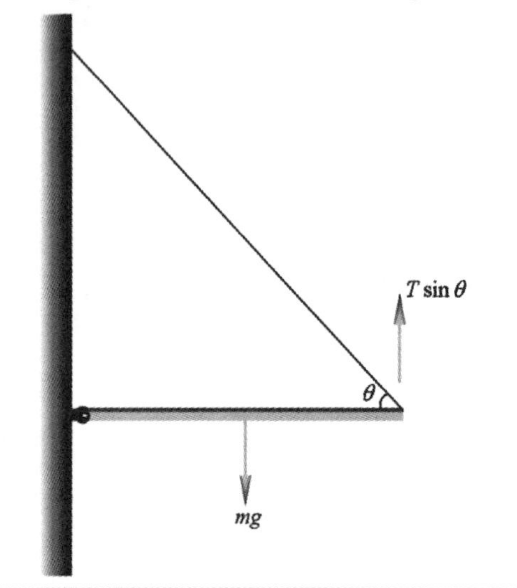

Figure 10.4

Review

Note that both sum-of-torques equations do not have sine terms attached to the weight of the strut. Again, this is due to the fact that sin 90° = 1

$$\sum \tau = 0$$

$$(T \sin 53.130412°)(3\,\text{m}) - (50\,\text{Kg})\left(9.8\frac{\text{m}}{\text{s}^2}\right)(1.5\,\text{m}) = 0$$

$$T(2.4000097) = 735\,\text{N} \cdot \text{m}$$

$$T = 306.2487622\,\text{N}$$

At 50°C

We will again use the sum-of-torques equation to find the tension in the cable. In this case, however, we must use the new angle that the cable makes with the strut, as well as the new center of mass.

$$\sum \tau = 0$$

$$(T \sin 53.100412°)(3.00324\,\text{m}) - (50\,\text{Kg})\left(9.8\frac{\text{m}}{\text{s}^2}\right)(1.50162\,\text{m}) = 0$$

$$T(2.401658) = 735.7938\,\text{N} \cdot \text{m}$$

$$T = 306.3691\,\text{N}$$

Conclusion

If desired, we could continue to compare the strut at the two temperatures and solve for both the horizontal and vertical components of the force on the strut due to the hinge. However, our purpose in this application was to show that thermal expansion is an effect that must be taken into account when doing structural and design analysis. In our simple example, we saw that the effect of a 45°C temperature difference generated an extremely small effect on both the angle between the cable and the strut and the tension in the cable. Thus, the level of accuracy required of the design project determines whether or not the designer must take thermal effects into account.

Exercises

1. Calculate the change in the length of a 1 m long, thin aluminum strut when the strut is raised by 40°C.

2. Using the coefficient of linear expansion for copper found in Table 10.1 of *College Physics*, find the coefficient of volumetric expansion for copper.

3. The volume of a copper sphere at 5°C is 0.5 m^3. Find the new volume of the sphere when the temperature is raised to 70°C.

4. If the aluminum strut in the figure is raised by 40°C, find the new center of mass of the strut measured from the left end of the strut.

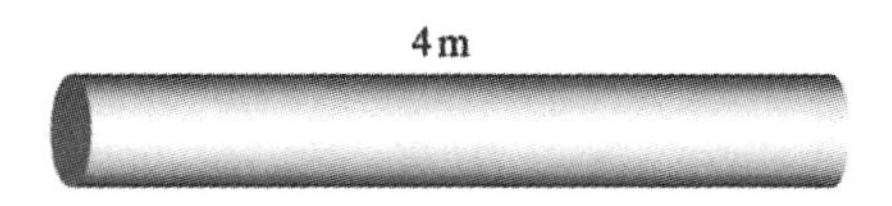

5. The 50 Kg steel strut in the figure is 4 m long when at 0°C.

   a) Find the tension in the support cable at this temperature.
   b) Recalculate the tension in the cable if the strut is raised to 35°C. Assume that only the strut experiences an increase in temperature.

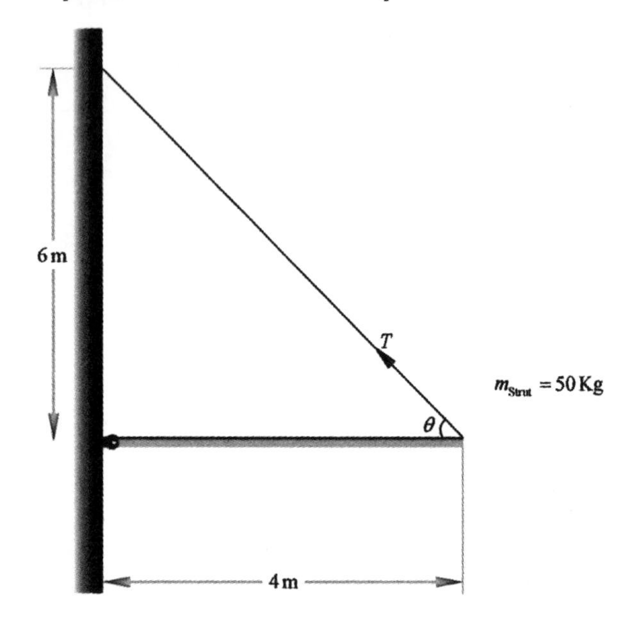

STRUCTURAL ANALYSIS AND DESIGN 11

Columns, Beams, and Trusses

In this application, we study the topic of stress as it applies to support columns, beams, and trusses.

Connection

A discussion of stress is found in Section 9.1 of *College Physics.*

In this application we explore the topic of stress. Stress is central to the analysis required to build stable structures and supports.

In Section 9.1 of *College Physics*, stress is introduced as a measure of a force causing a deformation. In equation form, stress is defined as the amount of deforming force applied to a cross-sectional area:

$$\text{stress} = \frac{F}{A}$$

Because forces can be applied in many different ways, we will examine several different types of stress.

Types of Stress

Tensile Stress and Compressional Stress

A deforming force, $\vec{F}$, can be applied to the ends of a rod so that it increases the length of the rod

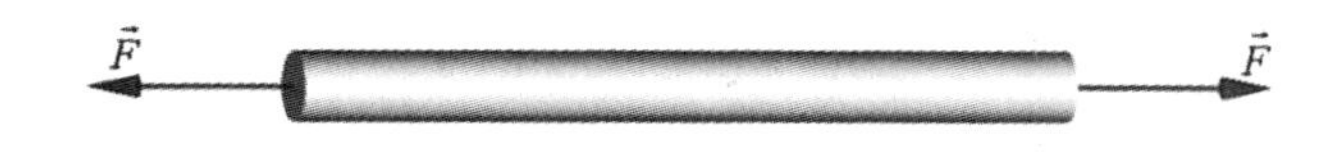

Figure 11.1

or decreases the length of the rod

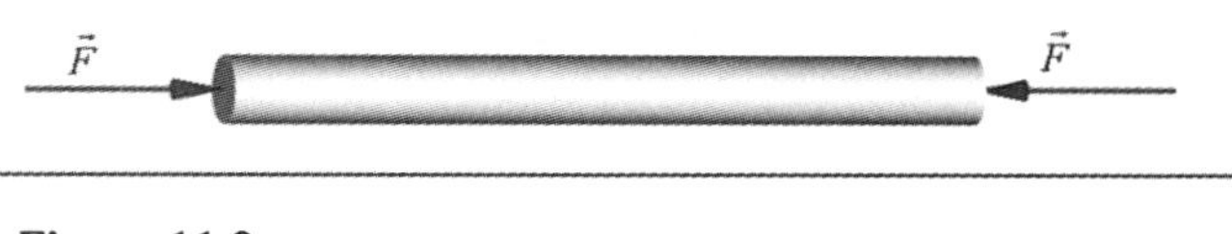

Figure 11.2

Because the force in Figure 11.1 places the rod under tension, this type of stress is referred to as *tensile stress*. Similarly, because the force in Figure 11.2 compresses the rod, this type of stress is called *compressional stress*. Notice that in both cases, the volume of the rod changes because of the applied stress.

Shear Stress

A deformation can also be caused by applying a force tangent to the surface area instead of perpendicular to it.

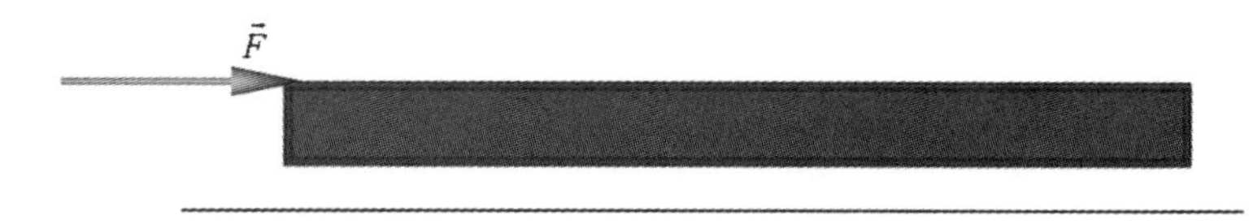

Figure 11.3

In this case, the stress is referred to as *shear stress*. Because the deforming force is tangent to the surface of the object, the object changes shape but does not change in volume.

Figure 11.4

Ultimate Strengths of Materials

For small deforming forces, the deformation of the object is proportional to the amount of applied force. Those forces that are small enough not to cause a permanent deformation correspond to the *elastic region* of the graph.

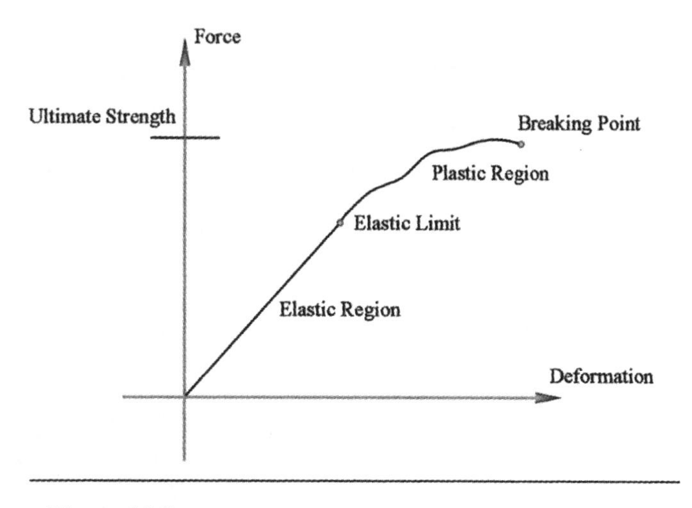

Figure 11.5

In this region, if the applied force is removed, the object will return to its original size and configuration. Once an object reaches its *elastic limit*, the object will not return to its original size even after the applied force is removed. Those forces that are large enough to cause a permanent deformation but not fracture the material correspond to the *plastic region* of the graph. If the applied force continues to increase, the object reaches its *breaking point*. The maximum force per unit area that a material can tolerate before fracturing is known as its *ultimate strength*. A listing of the ultimate strengths of a few materials is given in Table 11.1.

	Tensile Strength (N/m^2)	**Compressional Strength (N/m^2)**	**Shear Strength (N/m^2)**
Steel	500×10^6	500×10^6	250×10^6
Aluminum	200×10^6	200×10^6	200×10^6
Concrete	2×10^6	20×10^6	2×10^6
Wood (*Parallel to Grain*)	40×10^6	35×10^6	5×10^6

Table 11.1

Note that the ultimate strengths of the various materials are not uniform, but vary depending upon the type of stress experienced. Consequently, when making structural decisions, the type of stress that a material will experience must be known when choosing materials. Because the ultimate strength of a material depends on many factors, including impurities, it is standard to incorporate a safety factor of at least three, which means that we only assume an ultimate strength of 1/3 of the given table value for a material.

Stress-Bearing Design Elements

Support Columns

One of the most important things to consider when designing structures is the material to be used in the support columns. Because a structural column must support the downward-directed weight of the floors, roofing, etc., that lie above, the shear stress and tensile stress on the column are negligible when compared to the compressional stress. Thus we must choose a material that has a relatively high ultimate strength under conditions of compressional stress.

Courtesy of Prentice Hall

Figure 11.6

For aesthetic reasons or because of material availability, we may decide not to choose a material that has the absolute highest ultimate strength under compressional stress. When selecting a material with a smaller compressive strength, we can counter undue compressional stress by regulating the size of the column.

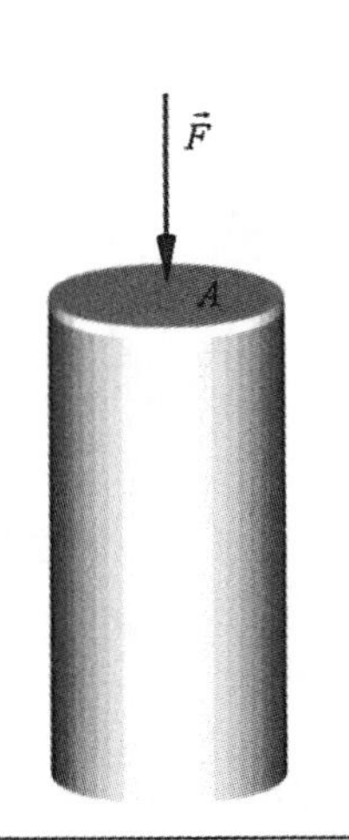

Figure 11.7

Because stress is defined as the ratio of the amount of deforming force to the cross-sectional area receiving the force

$$\text{stress} = \frac{F}{A}$$

the larger the cross-sectional area of the column, the smaller the stress. Thus, we only need to know the amount of downward force that the column(s) must support in order to choose the correct cross-sectional area for the column(s).

Example 1

What is the maximum force a circular concrete column of radius 0.5 m can support if a safety factor of 5 is used?

Solution The stress definition given can be rearranged and expressed as

$$F = (\text{stress}) \cdot A$$

The maximum compressional strength of concrete, the safety factor of 5, and the area of the column can then be inserted, yielding

$$f_{\text{max}} = \frac{1}{5}\left(20 \times 10^{6} \frac{\text{N}}{\text{m}^{2}}\right)\left[\pi (0.5\ \text{m})^{2}\right]$$

$$f_{\text{max}} = 3.14 \times 10^{6}\ \text{N}$$

By inserting a factor of 1/5 into the equation, we incorporate the safety factor of 5 and find that the column can support a maximum force of 3.14×10^{6} N.

Beams

The stress analysis for a horizontal beam is more complex because it is possible for the beam to experience all three types of stress simultaneously. A beam's intrinsic weight tends to cause the central portion to sag. Thus, while the top of the beam is under compressional stress, the bottom portion of the beam is under tensile stress. Other forces act internally to cause shear stress on the material.

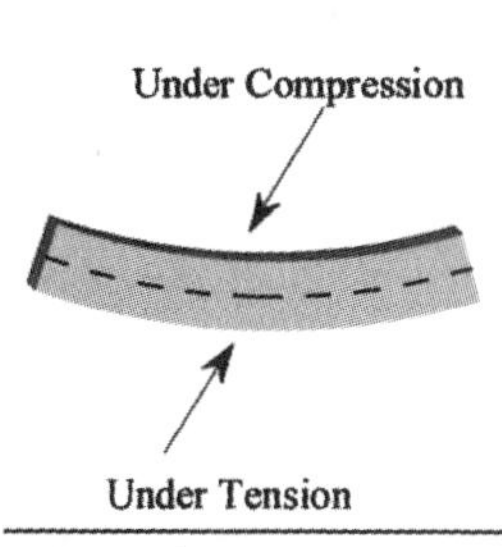

Figure 11.8

Internal reinforcement is often used to decrease the amount of deformation due to stress. For example, when constructing a highway, *reinforced concrete* is often used. This process involves lacing a grid of steel rods through the region in which the concrete is being poured. Although this process helps to reduce the deformation and stress, both compressional and tensile stress are still present as a result of the weight of the construction materials and the traffic that travels on the highway. Eventually, the concrete on the bottom of the highway will crack due to the tensile stress.

One effective means of countering the highway's fracturing is to use *prestressed concrete*. In this process, steel rods are again laced throughout the region in which the concrete is poured. However, in this case, the rods are placed under tension during the pour. The tension in the rods is released once the concrete has dried, which places the concrete under compression. This internal compression acts to balance the tensile stress that the bottom portion of the highway experiences due to its weight and the traffic.

Trusses

Connection

Trusses were briefly introduced in Structural Analysis and Design Application 4.

Because of the three stresses mentioned above, beams are ineffective architectural tools to use for spanning gaps that can range from the space between two walls that must support the roof of a building to the distance between two points on either side of a ravine. Instead, a *truss* is often used for this purpose.

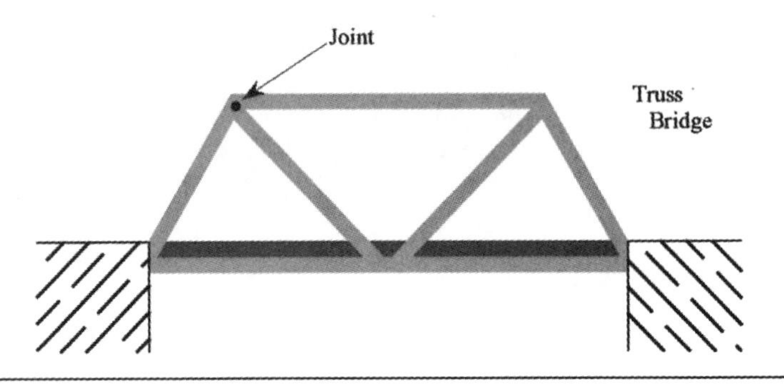

Figure 11.9

A truss is a system of struts that are linked together to form triangular patterns. The point at which two or more struts link together in the truss is called a *joint*. In a well-designed truss, when an external load is applied, each of the struts is placed under a purely compressional or a purely tensile stress. Thus, materials that are strong under these two types of stress, such as structural steel, must be used when constructing a truss.

FYI

It is important to note that by assuming that al of the forces in the trus act purely along the struts, we have created an idealized scenario that does not take into account the weight of the struts themselves. Even so, this approximation is valid in most cases that involve an analysis of the forces on a truss.

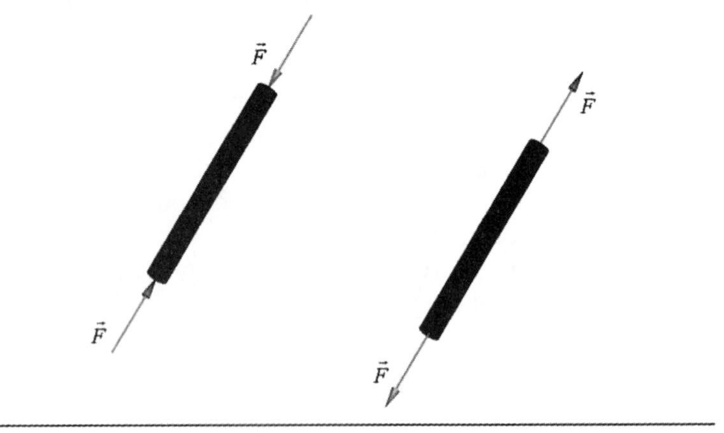

Figure 11.10

Conclusion

An understanding of the forces, tensions, and stresses involved in a design is crucial for the designer/drafter. These physical quantities affect facets of the design process ranging from the choice of materials to the appropriate style of architecture.

Exercises

1. Using a safety factor of 3, what is the maximum amount of weight that a circular concrete column can support if the column has a radius of 0.5 m?

2. At what point on a horizontal beam are the compressional and tensile stress each at their maximum value?

3. Discuss how the process of thermal expansion affects stress.

4. What tension must be applied to a circular strand of nylon with radius 0.5 mm in order to cause it to snap? Nylon has an ultimate tensile strength of 500×10^6 N/m^2.

5. Discuss how it is possible for a material to have a large compressional strength but a relatively small tensile stress.

6. Do the necessary research to find the design of each of the following. Make a sketch of each.

   a) The Warren Truss
   b) The Bowstring Truss
   c) The Pratt Truss
   d) The Scissors Truss

7. For each of the trusses listed in Exercise 6, give an example of a situation in which each might be used.

STRUCTURAL ANALYSIS AND DESIGN 12

Arches

In this short application, we use the concepts of force and stress to discuss the principles underlying the arches and domes used in many works of architecture.

Many works of architecture that we see today, from churches to bridges, use arches and domes as design and support elements. Both constructions date back to the time of the Roman Empire or earlier. Prior to the development of the engineering techniques that underlie arch and dome construction, the post and lintel structure was used to bridge gaps in space. A famous example of post and lintel construction is evident in the monumental structure of Stonehenge.

Figure 12.1

Although the post and lintel form of architecture was functional, the length of the horizontal top stone, or lintel, limited the size of the openings spanned. As discussed in Application 11, materials that are strong under compressional stress are not necessarily strong under tensile or shear stress. Because stone is not strong under tensile stress, the lintel cannot be very long or it will fracture. A solution to this architectural challenge was found in the form of the *arch*. In this application we explore the topic of stress as it relates to arches and domes. Stress is central to the analysis required to build stable structures that use arches as a design element.

Arches

In a well-crafted semi-circular arch, the forces exerted by the arch stones on each other cause mostly compressional stress.

Figure 12.2

The forces in the arch, both horizontal and vertical, are transferred to the side supports known as *buttresses*. In the case of a horizontal arch, the buttresses must provide a large amount of horizontal support to counter the stress imposed by the horizontal forces in the arch.

During the Gothic period of architectural history, designers discovered that they could decrease the amount of horizontal buttressing required if they constructed the arch with a pointed center rather than a rounded one.

Figure 12.3

The pointed center helps direct the stress more vertically on the supports. The downward direction of the weight vector and decrease in the amount of horizontal force due to the stones allows for a decrease in the amount of horizontal buttressing required from the supports. With the need for less horizontal buttressing, architects could be more creative in placing arch supports. Further, because a pointed arch can support more weight, they were freed to build heavier and more elaborate structures. We see one result of this creativity in the soaring ceilings and *flying buttresses* in many Gothic cathedrals, like the famous Notre Dame Cathedral of Paris.

Courtesy of Jean Robier, Paris

Figure 12.4

Domes

Domes are directly related to arches; in fact, the ceiling of a dome can be viewed as a continuous series of arches. The domes we find in churches and government buildings are essentially arches that have been rotated around a vertical axis.

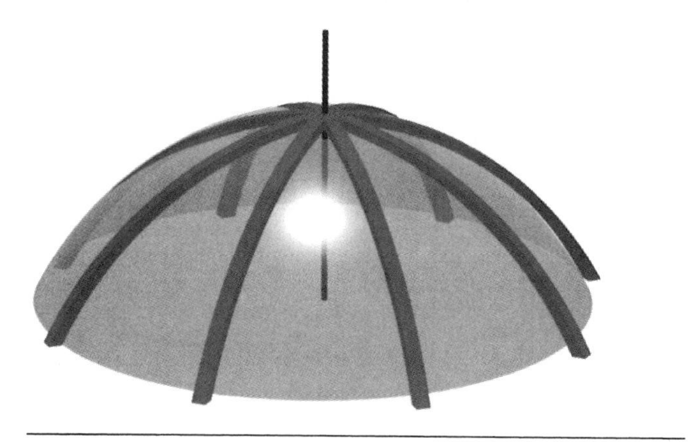

Figure 12.5

The same types of buttresses that are used in arch construction are placed around the base of the dome to counter the horizontal forces that are exerted along a line across the dome.

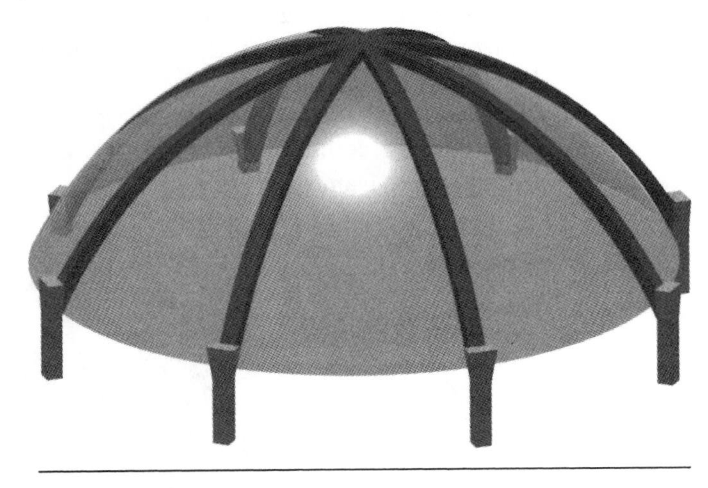

Figure 12.6

Exercises

In your region, find examples of the following types of architecture. If possible, discover the materials used in the construction of the structure and in the case of the arches and domes, discuss the buttressing used to support the structure.

a) Post and lintel
b) A round arch
c) A pointed arch
d) A dome

STRUCTURAL ANALYSIS AND DESIGN 13

Centripetal Force and the Inclination of a Banked Turn

In this application, we use an example of road construction taken from civil engineering to reinforce the topic of centripetal force.

Newton's First Law is found in Section 4.2 of *College Physics.*

Newton's First Law tells us that unless acted upon by a force, objects will move in straight lines and with constant velocities. Consequently, if an object or a system moves in a circular orbit, an external force must be causing the circular trajectory.

If an external force acting on a rotating object allows the speed of the object to remain constant, this force cannot have a component along the axis of the object's velocity. If it did, the speed of the object would change. As illustrated in Figure 13.1, we see that if the force is oriented at right angles to the velocity vector of the object the force changes the object's *direction* of travel, but not its *rate* of travel.

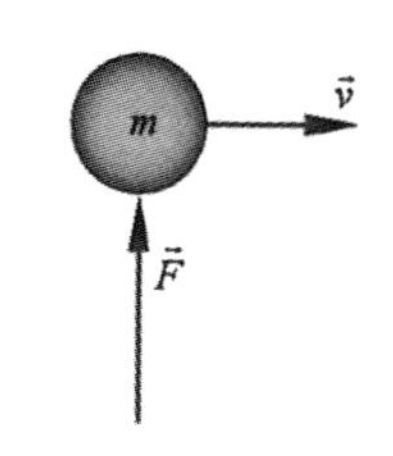

Figure 13.1

Centripetal Force

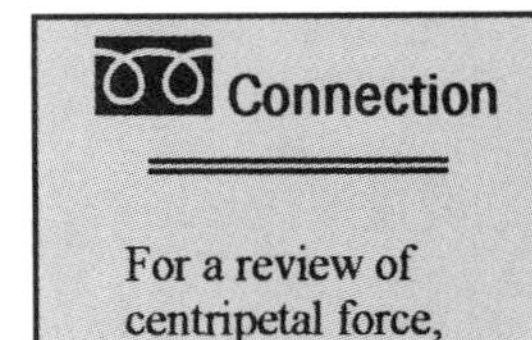

For a review of centripetal force, see Section 7.3 of *College Physics.*

An inward-directed force that changes the direction of the object so that it moves in a circular path is known as ***centripetal force*** (F_C).

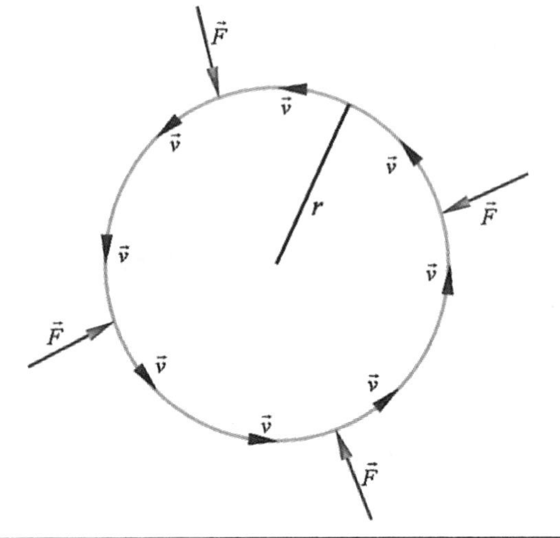

Figure 13.2

The Effect of a Centripetal Force

The circular orbit of an object or system is caused by an inward-directed centripetal force exerted on the object or system. This centripetal force changes the ***direction*** of the velocity vector, but not its magnitude.

An external force acting on an object or a system is required to provide this centripetal force. Many examples of this force are apparent in our physical world. In the case of a mass revolving on the end of a rope, it is the tension in the rope that provides the inward-directed centripetal force on the mass. Similarly, it is the gravitational force between the earth and the moon that provides the centripetal force that keeps the moon revolving in a circular orbit.

In this application, we look at a scenario drawn from civil engineering to examine the concept of centripetal force.

Example 1

Let's imagine that we are employed as an engineer. Our challenge is to determine the necessary angle of banking and the maximum speed at which a car may safely navigate a turn on a banked road we are designing. In this example, we assume that the road is frictionless.

> **FYI**
>
> Two examples of nearly frictionless roads are a road covered with wet leaves, and a road covered with ice.

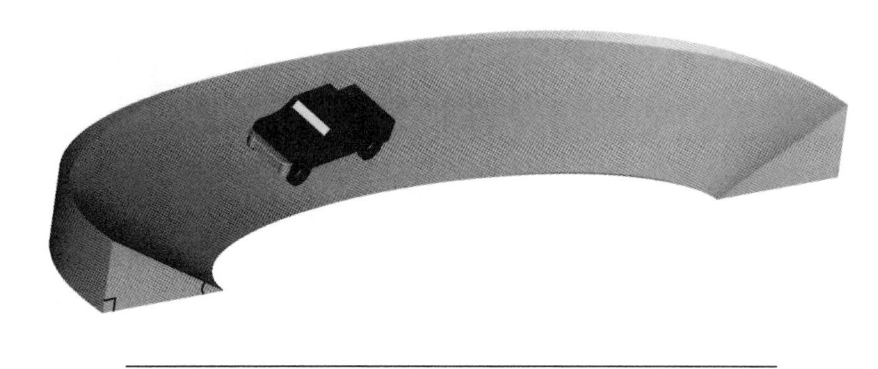

Figure 13.3

Solution Because the car takes the path of a circular orbit around the banked turn, a centripetal force is being exerted on the car. However, it is not readily apparent what causes this centripetal force.

To solve the problem, we draw a diagram of all of the forces acting on the car:

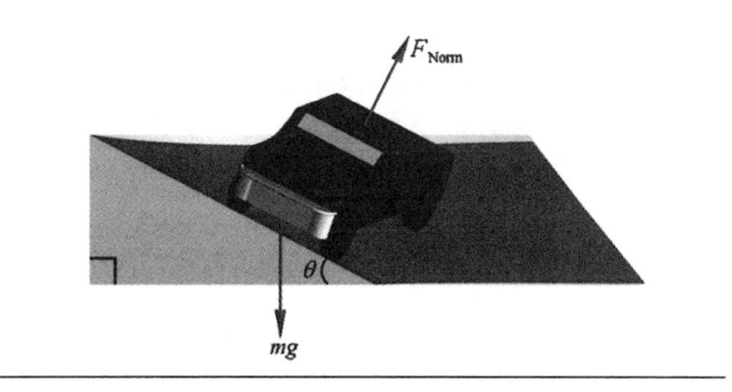

Figure 13.4

The normal force is discussed in Section 4.4 of *College Physics.*

Because we have excluded friction from this problem, the only inward-directed force that can provide the centripetal force is a component of the normal force exerted by the incline on the car:

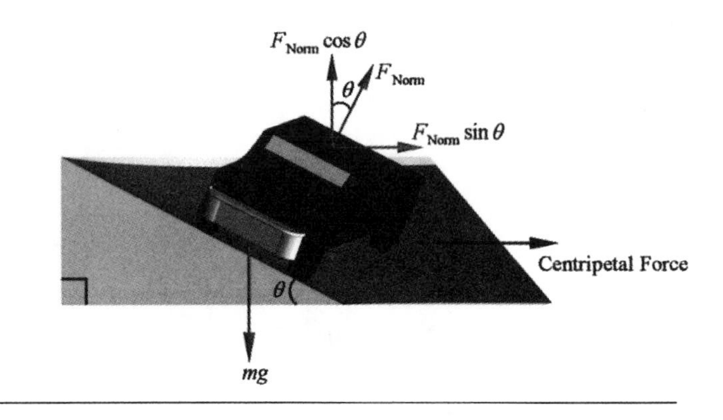

Figure 13.5

Once we realize that the horizontal component of the normal force is providing the centripetal force, we can use mathematics to find a solution to our design problem.

We begin by writing this information in the form of the equation

$$F_{Norm} \sin\theta = F_C \qquad \textbf{(1)}$$

Connection

Section 7.3 of *College Physics* presents the concept of centripetal force.

Recalling the formula given in *College Physics*, we know that the centripetal force on a mass (m) that is rotating with a velocity (v) in a circle of radius (r) can be written as

$$F_C = \frac{mv^2}{r}$$

Inserting this expression into Equation (1) gives us

$$F_{Norm} \sin\theta = \frac{mv^2}{r} \qquad \textbf{(2)}$$

Revisiting the force diagram in Figure 13.5, we see that because the car is moving neither upward nor downward, the two vertical forces on the car must be equal to one another. In equation form this is given by

$$F_{Norm} \cos\theta = mg$$

or,

$$F_{\text{Norm}} = \frac{mg}{\cos\theta}$$

Math Review

Since

$$\sin\theta = \frac{opp}{hyp}$$

and

$$\cos\theta = \frac{adj}{hyp}$$

if we divide the sine function by the cosine function, the result will be

$$\frac{\sin\theta}{\cos\theta} = \frac{opp}{adj}$$

which is the definition of the tangent function. Thus,

$$\frac{\sin\theta}{\cos\theta} = \tan\theta$$

If we insert this expression for the normal force into Equation (2), we get

$$\left(\frac{mg}{\cos\theta}\right)\sin\theta = \frac{mv^2}{r}$$

$$\frac{g\sin\theta}{\cos\theta} = \frac{v^2}{r}$$

$$g\left(\frac{\sin\theta}{\cos\theta}\right) = \frac{v^2}{r}$$

$$g\tan\theta = \frac{v^2}{r} \qquad \textbf{(3)}$$

Using Equation (3), we can now solve for the angle of banking (θ) or for the velocity of the car (v). To calculate the necessary angle of banking, we must know the velocity of the car. Alternatively, we can find the maximum safe velocity of the car if we know the angle at which the turn is banked.

Example 2

A car is attempting to round an icy bend in the road (negligible friction between the tires and the road). If the bend is banked at 20° and has a turn radius of 50 m, find the maximum velocity of the car that will allow it to negotiate the turn safely.

Solution Beginning with Equation (3) and solving for the velocity, we get

$$g\tan\theta = \frac{v^2}{r}$$

$$rg\tan\theta = v^2$$

$$\sqrt{rg\tan\theta} = v$$

Inserting the given values yields

$$v = \sqrt{rg\tan\theta}$$

$$v = \sqrt{(50\,\text{m})(9.8\,\text{m/s}^2)\tan 20°}$$

$$v = 13.35\,\text{m/s}$$

The maximum velocity that the car can travel to negotiate the icy bend safely is 13.35 m/s.

Conclusion

When designing roads with turns, civil drafters and designers must know the angle of banking required. In addition to determining the maximum safe speed for the turn, the amount of centripetal force that the normal force must provide determines the types of construction materials that must be used to build this portion of the highway.

Exercises

1. Calculate the centripetal force on a 2 Kg mass moving in a horizontal circle of radius 0.5 m with a velocity of 6 m/s.

2. A car is attempting to navigate a turn of radius 20 m that is banked at 10°. What is the maximum velocity at which the car can travel and still navigate the turn safely?

3. A car is moving at 55 mi/hr through a bend of radius 40 m.

   a) Using the methods of unit conversion, express the velocity of the car in m/s.
   b) At what angle must the turn be inclined in order for the car to navigate the turn?

4. A popular circus stunt involves an act in which a motorcyclist drives a motorcycle inside a horizontal hoop that is then raised above the crowd. Discuss how the concepts in this application apply to this stunt.

STRUCTURAL ANALYSIS AND DESIGN 14

Heating, Ventilation, Air-Conditioning, and the Continuity Equation

In this application, we use the Continuity Equation for fluids to discuss the heating, ventilation, and air-conditioning systems in a building.

Planning for the flow of air and ventilation throughout the building must be given careful consideration in the design of a structure. The number and size of vents in the various rooms must be calculated so that the building may be comfortably inhabited during both warm and cold seasons of the year. In this application, we discuss how to use the Continuity Equation from fluid dynamics to help analyze the heating, ventilation, and air-conditioning (HVAC) needs in a building.

The Continuity Equation

Connection

The Continuity Equation is discussed in Section 9.4 of *College Physics.*

Although we usually use the terms *fluid* and *liquid* interchangeably, they do not mean precisely the same thing. Both liquids *and gases* are classified as fluids. Thus, when discussing the Continuity Equation, it is important to remember that it can be used to analyze the flow of both liquids and gases.

The Continuity Equation relates the rate at which the velocity of a fluid flows through a section of a closed container to the cross-sectional area of the section. For example, if the fluid at the left end of the figure travels with a velocity of v_1, the velocity of the fluid cannot be the same at the right end because the cross-sectional area is smaller at that end.

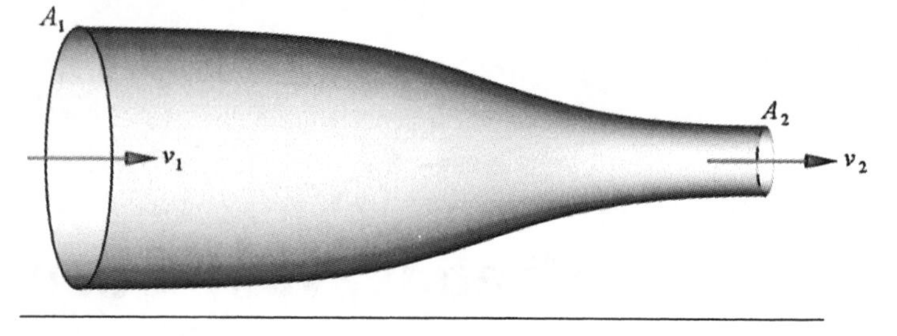

Figure 14.1

In fact, the velocity of the fluid at the right end must be higher so that the same amount of fluid can pass through both sections of the figure in the same amount of time. The equation that describes this relationship is given by the Continuity Equation

$$A_1 v_1 = A_2 v_2 \qquad \text{The Continuity Equation}$$

Example 1

Calculate the velocity of the fluid at the small end of the hose.

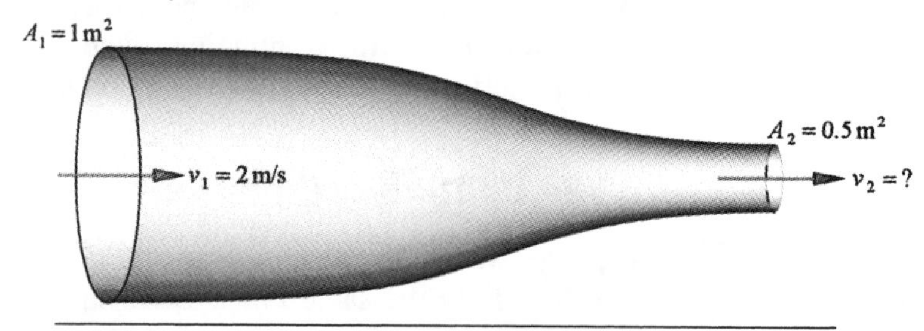

Figure 14.2

Solution To find a solution, we insert the given values for the area of the hose at each end along with the velocity at the left end into the Continuity Equation.

$$A_1 v_1 = A_2 v_2$$
$$(1\,\text{m}^2)(2\,\text{m/s}) = (0.5\,\text{m}^2)v_2$$
$$v_2 = 4\,\text{m/s}$$

Because the cross-sectional area at the right end is half the area at the left end, the velocity of the fluid at the right end is twice that of the left end.

HVAC

The Continuity Equation also applies to gases. Consequently it can be used to evaluate the airflow in a building. From the previous example, we know that if the air in the vent in Figure 14.3 moves with velocity v_1, it will not have the same velocity once it enters the room.

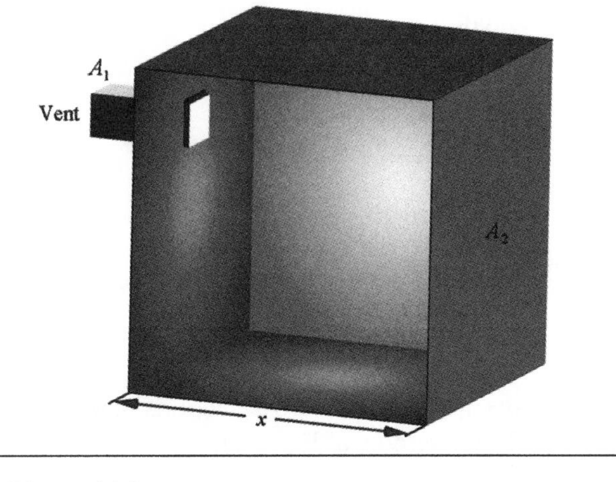

Figure 14.3

Because the cross-sectional area of the room is much larger than the cross-sectional area of the vent, the velocity of the air will be much smaller once it enters the room. Although the velocity of the warm or cool air moving through a vent might be quite high, this velocity cannot be used to calculate the amount of time required to heat or cool the air in the room.

Example 2

Find the cross-sectional area of a heating duct if the warm air moving through it at 4 m/s refreshes all the air in the room in 12 minutes.

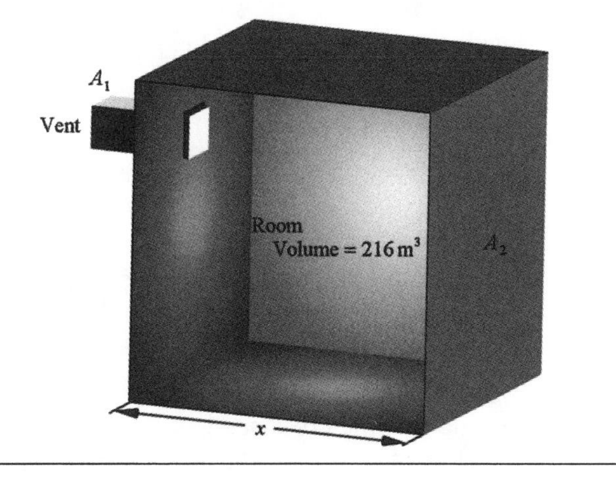

Figure 14.4

Solution The velocity of the air in the room can be expressed as the distance x divided by the time t that it takes the air to move across the room. Incorporating this information into the Continuity Equation,

$$A_1 v_1 = A_2 v_2$$

yields

$$A_1 v_1 = A_2 \cdot \frac{x}{t}$$

which can also be expressed as

$$A_1 v_1 = \frac{A_2 x}{t} \qquad \textbf{(1)}$$

Because A_2 is the cross-sectional area of the room and x is the length of the room, the numerator of Equation (1) is the volume of the room that is given as 216 m^3.

$$A_1 v_1 = \frac{216\,\text{m}^3}{t} \qquad \textbf{(2)}$$

The problem statement tells us that the air in the room is refreshed every 12 minutes. By converting to units of seconds, we find that it takes 720 seconds for the air to move across the room. Thus, Equation (2) becomes

$$A_1 v_1 = \frac{216\,\text{m}^3}{720\,\text{s}}$$

Finally, by inserting the value for the velocity of the warm air in the duct, v_1, and solving for A_1

$$A_1(4\,\text{m/s}) = \frac{216\,\text{m}^3}{720\,\text{s}}$$

$$A_1 = \frac{216\,\text{m}^3}{(4\,\text{m/s})(720\,\text{s})}$$

$$A_1 = 0.075\,\text{m}^2$$

we find that the cross-sectional area of the heating duct must be 0.075 m^2. It is interesting to note that the actual shape of the duct is irrelevant. Provided that the duct has a cross-sectional area of 0.075 m^2, it does not matter if the duct is circular, rectangular, or another shape.

Conclusion

When designing a civil structure, one must consider the rate at which the warm/cool air in the ventilation system is being generated as well as the number and size of rooms that must be ventilated. Inserting this information into the Continuity Equation then provides a platform from which a designer can predict the flow of air throughout the structure.

Exercises

1. Find the velocity of the fluid at the narrow end of the figure.

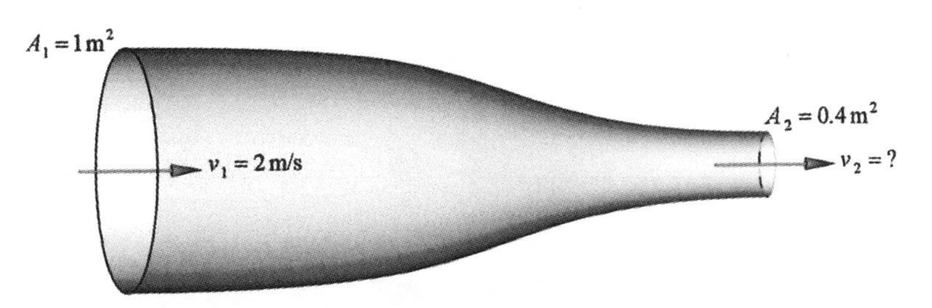

2. Find the velocity of the fluid at the wide end of the figure.

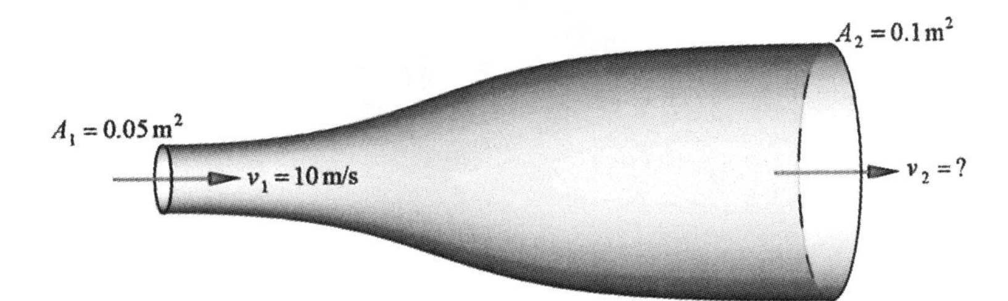

3. Find the velocity of the fluid at A_2 and A_3.

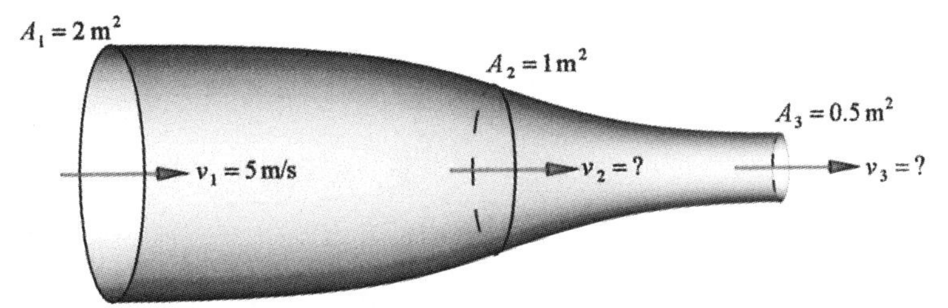

4. If the air from the vent refreshes all of the air in the room in 15 minutes, find the necessary cross-sectional area of the vent.

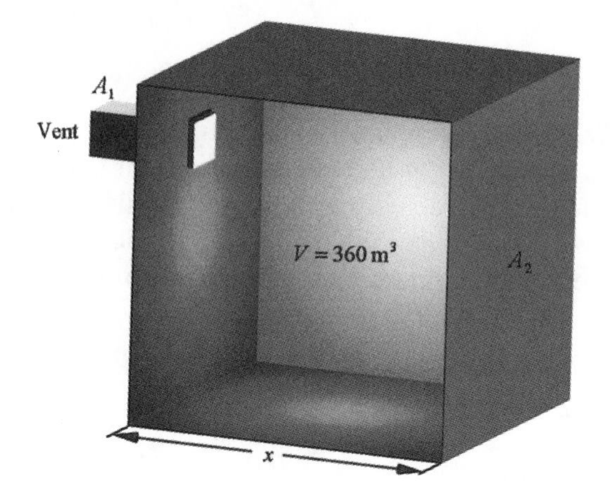

5. For each of the rooms, find the approximate amount of time required for the vents to refresh the air. Assume that the airflow from the 0.2 m^2 vent is split evenly between the rooms.

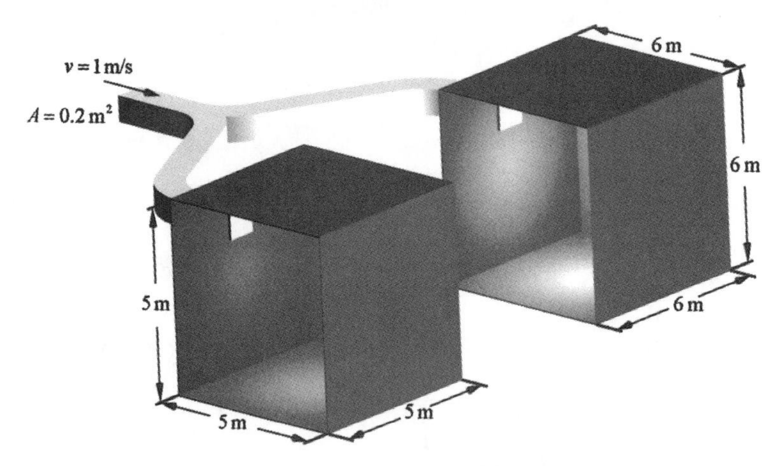

STRUCTURAL ANALYSIS AND DESIGN 15

The Foucault Pendulum

In this application, we extend the pendulum concepts introduced briefly in *College Physics* to a design element used in many modern structures, the Foucault Pendulum.

In the entrance to many impressive buildings, a large pendulum known as a Foucault Pendulum can be seen decorating the lobby. This pendulum oscillates back and forth inside a circle that is inscribed on the floor. If watched for a long enough time period, the pendulum appears to precess around the circle. In other words, the pendulum appears to "walk" around the inscribed circle as the pendulum swings back and forth. To gain a deeper understanding of the physical properties of this large-scale structural decoration, we will begin by discussing the simple pendulum and then examine the Foucault Pendulum itself.

The Simple Pendulum

Connection

Pendulum concepts are introduced briefly in Section 12.2 of *College Physics.*

Suppose that a mass is suspended from a cord of negligible mass, as illustrated in Figure 15.1

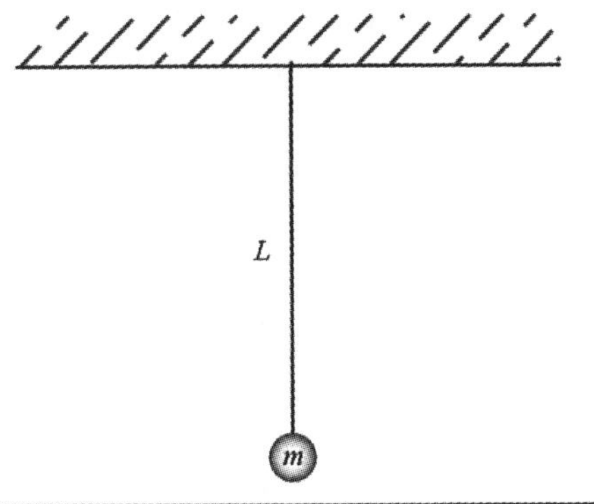

Figure 15.1

If the pendulum is displaced through a small angle, θ, a component of the weight of the mass will act to restore the mass to its equilibrium position.

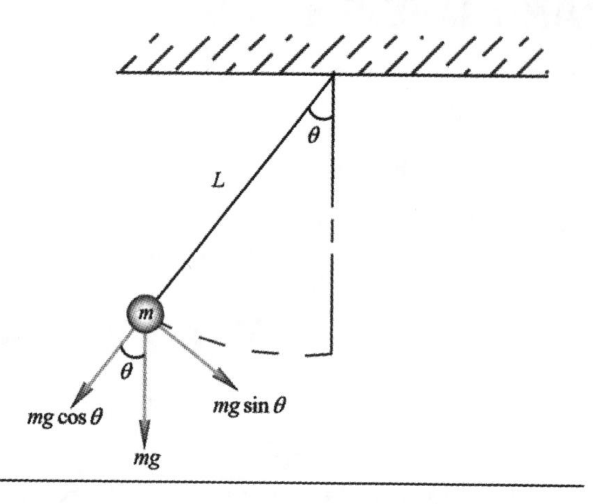

Figure 15.2

Because the restoring force acts opposite to the direction of the mass's displacement, we can express the restoring force as

$$F = -mg\sin\theta \qquad \textbf{(1)}$$

The Small-Angle Approximation

For small angular displacements, it is possible to replace $\sin\theta$ with the angle θ itself. Provided the angle is measured in radians, this replacement is reasonable because the difference is negligible. For example, for the angle 0.0653 rad, note the difference between

$$\theta = 0.0653\text{ rad} \qquad \text{and} \qquad \sin\theta = 0.0652$$

Because we assumed that our pendulum was displaced through a small angle, Equation (1)

$$F = -mg\sin\theta$$

can be rewritten as

$$F = -mg\theta \qquad \textbf{(2)}$$

The motion of the pendulum can be thought of as a section of a circle with radius L and arc length S.

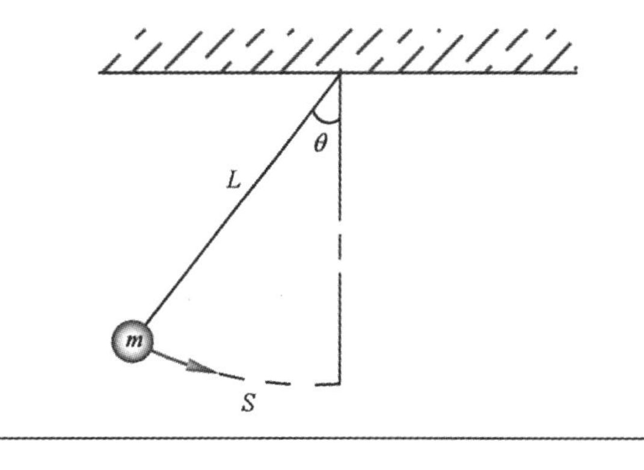

Figure 15.3

Therefore, the angular displacement, θ, in Equation (2) can be replaced using the relationship between the angle and the arc length familiar from geometry:

$$S = L\theta$$

Thus, Equation (2)

$$F = -mg\theta$$

becomes

$$F = -mg\left(\frac{S}{L}\right)$$

or

$$F = -\left(\frac{mg}{L}\right)S \qquad \textbf{(3)}$$

Connection

Hooke's Law is covered in Section 5.2 of *College Physics.*

If we now compare Equation (3) with Hooke's Law,

$$F = -kx$$

we see that they are strikingly similar. Both equations have a force that is proportional to, and in the opposite direction of, the displacement from equilibrium. In Hooke's Law, this displacement is recorded using the variable x, while in the equation for the pendulum the displacement is recorded using the arc length, S.

Connection

The period of a simple harmonic oscillator is discussed in Section 12.2 of *College Physics.*

Thus, we may interpret the term $\frac{mg}{L}$ in Equation (3) as being analogous to the force constant k in Hooke's Law. Given this interpretation, we can find an expression for the period of the pendulum using Equation 12.11 from *College Physics*. This equation is used to find the period of a mass-spring system.

$$T = 2\pi\sqrt{\frac{m}{k}} \qquad \textbf{(12.11)}$$

Connection

The fact that the period is independent of the mass can also be motivated using conservation of energy. The period of the pendulum is independent of the mass in the same way that the mass of a falling object does not affect the time that it takes the mass to hit the ground.

See Section 5.5 of *College Physics* for a discussion of freefall using conservation of energy.

Replacing k with its pendulum equivalent, $\frac{mg}{L}$, yields

$$T = 2\pi\sqrt{\frac{m}{\frac{mg}{L}}}$$

$$T = 2\pi\sqrt{\frac{m}{1} \cdot \frac{L}{mg}}$$

$$T = 2\pi\sqrt{\frac{L}{g}} \qquad \textbf{(12.14)}$$

which is Equation 12.14 from *College Physics*. Notice that the period is independent of both the mass and the amplitude of the oscillation. The time required for the pendulum to execute one full oscillation depends only upon the length of the pendulum and the local gravitational acceleration.

Example 1

Find the period of a simple pendulum that has a length of 1 m.

Solution By inserting the length and the value of the earth's gravitational acceleration into the period expression, we find

$$T = 2\pi\sqrt{\frac{1\,\text{m}}{9.8\,\text{m/s}^2}}$$

$$T = 2.0\,\text{s}$$

The period of a simple pendulum with a length of 1 m is 2.0 s.

The Foucault Pendulum

In 1851, Jean Bernard Foucault showed that a pendulum could be used to prove that the earth rotates on an axis. The Foucault Pendulum, named after this brilliant scientist, typically consists of a pendulum that hangs above a circle that is inscribed on the floor. This circle corresponds to the endpoints of the pendulum's oscillation. As the pendulum executes simple harmonic motion along its plane, the earth rotates beneath the pendulum. Because of this planetary rotation, the pendulum appears to precess around the circle drawn on the floor.

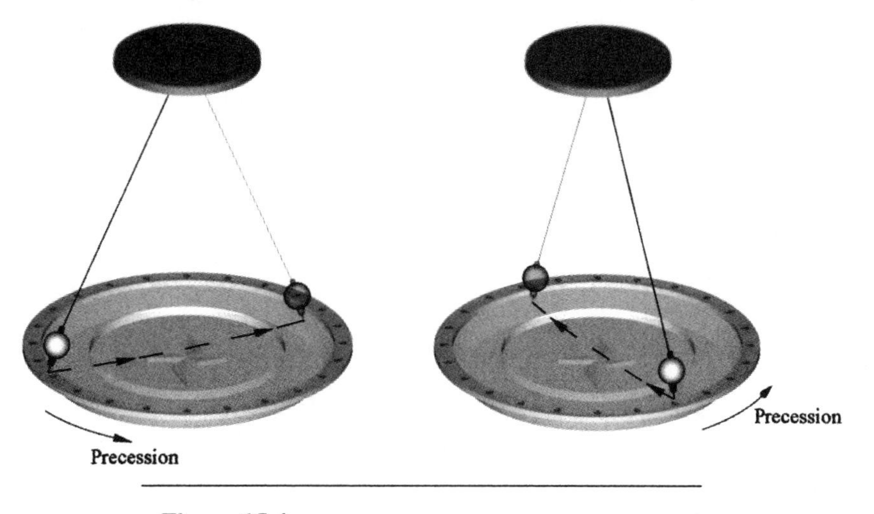

Figure 15.4

Conclusion

When incorporating decorative elements into designs, it is often important to understand the basic physical principles of the design element so that it works with the rest of the design. In this application, this knowledge takes the form of understanding the period of a simple pendulum so that the appropriate size of pendulum is used in the building. However, in other designs, this knowledge can take the form of the amount of friction between the wall decorations and a wall, the tension in the cables that support a decorative form of lighting, or a variety of other possibilities.

Exercises

1. Calculate the period of a 2 m pendulum.

2. If a simple pendulum has a period of 0.5 sec, find the length of the pendulum.

3. Calculate the period of a 1 m pendulum on the moon. Assume that the moon has a gravitational acceleration one-sixth that of earth's.

4. Find the length and circle radius of a Foucault Pendulum located at the North Pole that makes one full transit of the inscribed circle every twenty-four hours.

5. What value of spring constant would be required for the mass-spring system and the pendulum to have the same period?

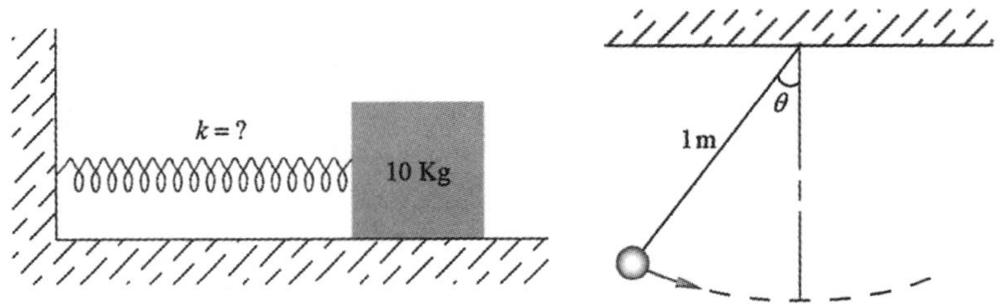

$k = ?$

10 Kg

1m

θ

Part II: Electronics

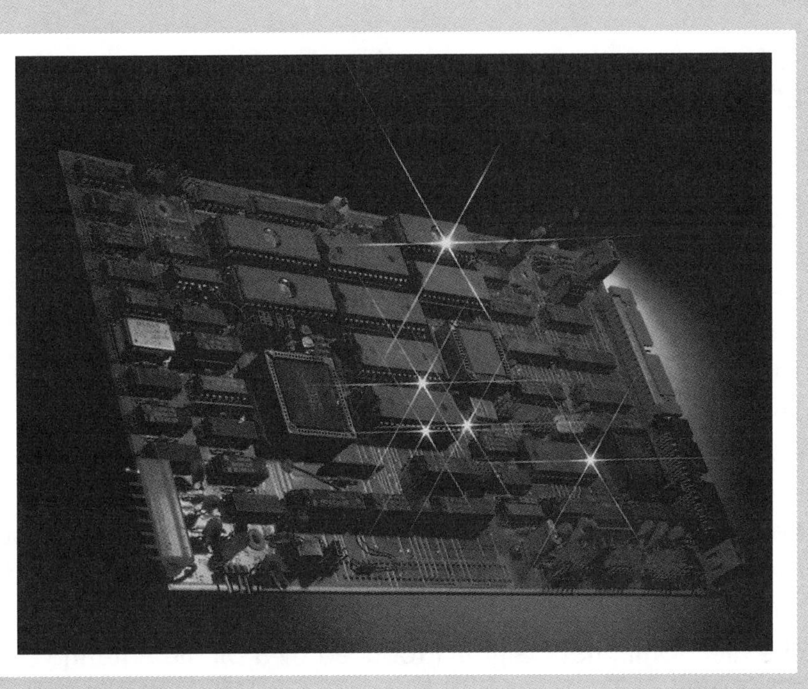

Courtesy of Photo Disc, Inc.

This part contains fifteen applications that illustrate how the language and concepts of physics apply to the field of electronics. These applications demonstrate how gaining an understanding and appreciation of physics helps to deepen our knowledge of electronics.

The section opens with applications that address the elementary topics of significant figures and units. The core group of applications, ranging from a discussion of voltage to the thermal behavior of semiconductors, explores familiar electronics settings in greater depth using the language and principles of physics. The final two applications show how the physics of the 20th and 21st centuries are influencing the next generation of technology.

Part II: Electronics

Applications

1. Significant Figures and the Charge on the Electron
2. Unit Analysis and Conversions in Electronics
3. Density and Electronics
4. Vectors and Electronics
5. The Force on an Electron in an Electric Field
6. The Force Between Two Current-Carrying Wires
7. A Comparison of the Gravitational and Electromagnetic Forces
8. Potential Versus Potential Energy
9. Applying the Principles of Energy to Series and Parallel Resistors
10. Heat, Kinetic Energy, and the Breakdown of PN-Junction Diodes
11. The Power Dissipated by an RLC-Circuit
12. The Electromagnetic Signal Produced by a Dipole Antenna
13. Diffraction and Television Signals
14. Relativistic Kinetic Energy and Momentum
15. Physics and the Next Generation of Computing

ELECTRONICS 1

Significant Figures and the Charge on the Electron

In this brief application, we use the amount of electric charge possessed by an electron as an illustration of *significant figures.*

Connection

The rules of significant figures are discussed in Section 1.6 of *College Physics.*

J.J. Thompson

Courtesy of Library of Congress

The rules of significant figures provide a method for attaching a measurement of both error and importance to numbers. Simply stated, the least precise measurement involved in a calculation limits the precision of the numbers and predictions of the calculation. In this application, we will use the electronics example of the charge on an electron to acquire an understanding of the usefulness of this mathematical tool.

The discovery of the electron is credited to J.J. Thompson, who carried out a series of experiments between 1894 and 1899 at Cambridge University. This series of experiments proved that the highly controversial world of the very small actually existed. In addition to discovering the existence of the electron, Thompson's experiments found an approximate value for its charge. A precise value for the charge was found later by the American, Robert Millikan, and his graduate student H. Fletcher using a painstaking experiment that measured the amount of electric charge on individual drops of oil.

Today, the charge on the electron is known to great accuracy and is given by

$$q_e = -1.60217733 \times 10^{-19}\ \text{C}$$

However, in almost all electronics calculations, the charge is abbreviated to

$$q_e = -1.6 \times 10^{-19}\ \text{C}$$

The question then arises, why do we use this less-precise value if the actual value is known? The answer lies in the principles of significant figures. Recall that basic electronics involves values such as 10 Volts, 4.7 KΩ, etc. Because for most measurements of electronics quantities we do not use more than one- or two-decimal-place accuracy, it does not make sense to use eight-decimal-place accuracy for the charge on the electron.

FYI

The *tolerance* of resistors also plays a role in determining the number of significant figures used in resistor calculations.

Example 1

A calculation of the current flow, I, using the more precise value of the electric charge is pointless if it is then multiplied by the less precise resistance of 4.7 KΩ in the equation for Ohm's Law,

$$V = IR$$

The precision of the calculated voltage is determined by the precision to which we know the resistance.

Example 2

When doing electronic signal calculations, the speed of electromagnetic waves such as radio waves in a vacuum is given by the precise value

$$c = 2.99792458 \times 10^8 \text{ m/s}$$

Connection

The speed of light is also discussed in Electronics Application 12.

Again, because of the precision of the other electronic quantities that are used in these types of calculations, the speed of the waves is usually rounded to

$$c = 3 \times 10^8 \text{ m/s}$$

Conclusion

When executing physical calculations, like those in electronics, it is important to know the accuracy of the measurements used in the calculations. The rules of significant figures presented in *College Physics*, and discussed briefly in this application, provide a method for determining the accuracy of physical measurements and calculations. Because of the level of accuracy required in today's technology, the ability to use and understand significant figures is a necessary tool for anyone entering the field of technology.

Exercises

1. From your previous electronics studies, identify which types of electronics measurements are made with more than one decimal place of accuracy. Which measurements do not allow for this degree of accuracy?

2. When dealing with common-emitter transistor circuits, we are often able to make the approximation that the current through the collector is the same as the current through the emitter:

$$I_C \cong I_E$$

   Use the method of significant figures to discuss under what conditions this assumption is permissible.

3. List and discuss any electronics situations in which it would be appropriate to use a more precise version of the electronic charge than

$$q_e = -1.6 \times 10^{-19}\ \text{C}$$

ELECTRONICS 2

Unit Analysis and Conversions in Electronics

In this application we use examples drawn from electronics to illustrate the methods of unit conversions and unit analysis introduced in Chapter 1 of *College Physics.*

The *units* attached to a measurement are labels that identify the type of quantity being measured. For example, the expression 5 seconds is a measurement of time, while 5 inches is a measurement of length. Although both measurements have a numerical value of 5, it is the unit label of seconds and inches that identifies the physical quantity being measured.

As discussed in Section 1.2 of Wilson/Buffa's *College Physics*, a number of different systems of measurement with different units are used throughout the world. Because different people find different systems of units preferable, it is important to be able to translate measurements from one system to another. This translation is called a *unit conversion.* It is important to note that when we perform a unit conversion, we do not change the size of the measurement, only the units in which the measurement is expressed.

Unit Analysis and The RC-Time Constant

Connection

Unit analysis is discussed in Section 1.4 of *College Physics.*

Unit analysis is a powerful tool that helps us gain physical information about quantities not contained in the numerical values given in a problem. In this section, we will examine a circuit containing a resistor, a capacitor, and a direct current voltage source to illustrate the importance of analyzing the units in a problem. Specifically, we will see how unit analysis helps to explain the name and meaning behind one of the familiar quantities from basic electronics, the *RC-Time constant.*

If the switch is closed in the circuit in Figure 2.1, the capacitor will begin to charge.

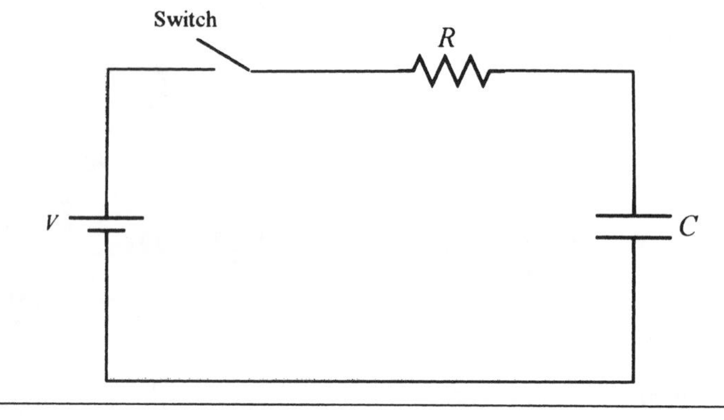

Figure 2.1

The amount of charge on the capacitor, q, as time passes is found using the equation

Math Review

Note that due to the presence of the exponential function in the charge equation, the capacitor approaches its maximum value in an exponential rather than a linear manner.

$$q = CV\left(1 - e^{-\frac{t}{RC}}\right) \qquad \textbf{(1)}$$

Recall from previous electronics courses that the product RC in this equation has a specific meaning. Referred to as the *RC - Time Constant*, this quantity is the amount of time required for the capacitor to reach 63% of its maximum charge. Further, at five such time constants, the capacitor reaches a charge close enough to its maximum value that a technician can treat the capacitor as being fully charged. In this application, we will explore why it is possible to treat the product of two *electronics* quantities as an amount of *time*. Unit analysis will provide the answer to this question.

The Units on the RC - Time Constant

To begin, let's look at the units on each of the quantities R and C. From earlier courses, we know that resistance is measured in ohms and capacitance is measured in farads. We can, however, use different units to describe these two quantities.

To attach a different unit to resistance we use the relationship

$$V = IR$$

and rearrange it so that

$$\frac{V}{I} = R$$

This relationship tells us that the units of resistance can also be expressed as the units of voltage divided by the units of current. Thus, instead of using ohms to measure a resistance, we can use volts/amps.

In a similar manner, we can use units other than farads to measure capacitance.

The equation

$$C = \frac{q}{V}$$

defines the relationship between the charge on the capacitor, the capacitance, and the voltage across the capacitor, and indicates that we can also use coulombs/volts as units of capacitance.

Using these alternative units, the units on the product *RC* become

$$\left[\frac{\text{volts}}{\text{amps}}\right]\left[\frac{\text{coulombs}}{\text{volts}}\right]$$

or when we cancel out the volts,

$$\left[\frac{\text{coulombs}}{\text{amps}}\right]$$

The definition of an ampere is one coulomb of charge flowing past a point per second; therefore, we can rewrite the units as

$$\left[\frac{\text{coulombs}}{\dfrac{\text{coulombs}}{\text{sec}}}\right]$$

or

$$\left[\frac{\text{coulombs}}{1} \cdot \frac{\text{sec}}{\text{coulombs}}\right] = [\text{sec}]$$

Using alternative units, we see that the unit for the product RC is indeed a unit of time. Because the unit is seconds it is reasonable to identify this product as a time constant and to make references to the amount of time corresponding to, for example, 5 time constants.

An Alternative Method for Finding the Units on the RC-Time Constant
In closing this section, we note that there is also an intuitive method of examining the units in this problem. Since it is not possible for the exponent on a number to have units, the units must cancel in the term $\frac{t}{RC}$. Thus, since time is measured in seconds, the denominator RC must also be measured in seconds.

Unit Conversions and Electron-Volts

Connection

The SI standard system of units is discussed in Sections 1.2 and 1.3 of *College Physics.*

In this section, we will review how to do a unit conversion from the macroscopic energy of joules, used for transformers and other large-energy electronics settings, to one more appropriate for the individual electrons that comprise an electric current, the *electron-volt* (eV).

Although it is possible to carry out measurements of electrons using joules as the energy unit, the large scale of this unit of measure makes it inconvenient. The unit of electron-volts is a more appropriate unit for this microscopic scale.

This energy unit is the amount of energy that an electron acquires as it moves through a one-volt potential difference. The relationship between the macroscopic energy unit of joules and the microscopic unit of electron-volts is given by

Connection

We will discuss the unit *electron-volts* in greater depth in Electronics Application 8 - *Potential Versus Potential Energy.*

$$1\,\text{eV} = 1.6\times10^{-19}\,\text{J} \qquad \textbf{(2)}$$

The expression in Equation (2) is known as a *conversion factor*. Conversion factors are used to translate a measurement from one system of units to another. Note that Equation (2) is just a mathematical way of stating that 1 eV and 1.6×10^{-19} J are the same amount of energy. A listing of many of the more useful conversion factors is found on the inside front cover of Wilson/Buffa's *College Physics* and this text.

Example 1
Express the energy measurement of 500 eV in joules.

Solution To solve this problem, we use the given conversion factor in Equation (2) to change the units on the measurement. Because our purpose is to change the units on the measurement but not the *size* of the measurement, we must be careful in selecting the mathematical operations that we can perform on the measurement.

Notice that if we take Equation (2) and divide both sides by 1 eV, we are left with 1 on the left side:

$$1\,\text{eV} = 1.6\times10^{-19}\,\text{J}$$

$$\frac{1\,\text{eV}}{1\,\text{eV}} = \frac{1.6\times10^{-19}\,\text{J}}{1\,\text{eV}}$$

$$1 = \frac{1.6\times10^{-19}\,\text{J}}{1\,\text{eV}} \qquad \textbf{(3)}$$

Rewriting the expression in the form of Equation (3) is helpful because multiplication by 1 is a mathematical operation that does not change the size of a measurement. Thus, if we multiply a measurement by the right side of Equation (3), we will change the units on the measurement, but not its size.

Our unit conversion then takes the form

$$500\,\text{eV}\cdot\frac{1.6\times10^{-19}\,\text{J}}{1\,\text{eV}}$$

Notice that the original units of electron-volts cancel between the measurement and the conversion factor, leaving joules as the result.

Thus,

$$500\,\text{eV}\cdot\frac{1.6\times10^{-19}\,\text{J}}{1\,\text{eV}} = 500(1.6\times10^{-19}\,\text{J}) = 8.0\times10^{-17}\,\text{J}$$

Five hundred electron-volts of energy is the same amount of energy as 8.0×10^{-17} joules.

Example 2

Express the energy measurement of 0.53 J in electron-volts.

Solution To convert from joules to electron-volts, we need to rearrange Equation (2) so that the unit of joules is in the denominator of the conversion factor. This placement will allow us to cancel the units of joules on the measurement. To achieve this result, we divide both sides of Equation (2) by 1.6×10^{-19} joules:

$$\frac{1\,\text{eV}}{1.6\times10^{-19}\,\text{J}} = 1$$

Our conversion then becomes

$$0.53\,\text{J}\cdot\frac{1\,\text{eV}}{1.6\times10^{-19}\,\text{J}}=\frac{0.53(1\,\text{eV})}{1.6\times10^{-19}}=3.3\times10^{18}\,\text{eV}$$

0.53 joules is the same amount of energy as 3.3 x 10^{18} electron-volts

Conclusion

An understanding of the units in a calculation can give an engineer or technician additional information that is not contained in the numerical portion of the calculation. In this application, we were able to explain the *RC*-time constant from dc electronics by analyzing the units involved. In addition, the units in a calculation provide a mechanism by which we can move from the energies of the individual electrons that comprise an electric current to the energies involved with the electronic devices that are used daily in electronics laboratories.

Exercises

1. Convert the energy measurement of 600 eV into joules.

2. Convert the energy measurement of 2 J into electron-volts.

3. The amount of potential energy, expressed in joules, stored inside an inductor is given by

$$\text{PE}=\frac{1}{2}Li^2$$

   a) Find the amount of potential energy stored inside a 4 μH inductor if there are 4.5 mA of current flowing through the inductor.
   b) Express your answer in electron-volts.

4. The amount of potential energy, expressed in joules, stored inside a capacitor is given by

$$\text{PE}=\frac{1}{2}CV^2$$

   a) Find the amount of potential energy stored inside a 5 μF capacitor that has 6 volts across its plates.
   b) Express your answer in electron-volts

5. In today's particle accelerators, it is not uncommon for scientists to accelerate electrons into the gigaelectron-volt (GeV) range. What is the joule equivalent of an electron that possesses 1 GeV of energy?

ELECTRONICS 3

Density and Electronics

In this application, we apply the concept of density to two electronics situations: the surface charge density of a parallel-plate capacitor, and the current density in a wire.

> **Connection**
>
> The equation for density is introduced in Section 1.4 of *College Physics.*

The Concept of Density

In *College Physics*, the physical quantity of *density* is defined as being the ratio of the mass of an object to the amount of three-dimensional space, or volume, occupied by the object, and is expressed by the equation

$$\rho = \frac{m}{V}$$

Example 1
Find the density of a stone that has a mass of 10 Kg and occupies a volume of 2 m^3.

Solution To find the density of this object, we simply insert the given values of the mass and volume into the density expression and simplify:

$$\rho = \frac{m}{V}$$

$$\rho = \frac{10\,\text{Kg}}{2\,\text{m}^3}$$

$$\rho = 5\frac{\text{Kg}}{\text{m}^3}$$

Extending the Concept of Density

By extending this simple definition to allow for a broader interpretation of the concept of density, we can see how it is applicable to the field of electronics.

In addition to thinking of density as simply being mass divided by volume, we can interpret it much more broadly. For example, if we compare the number of people who live in a five-square-mile region of rural Montana to the number of people who live in the same size area in the Silicon Valley of California, we see that the Silicon Valley has a much higher *population density*. In this case, the density is expressed as the number of people per square mile.

As another example, if we look at the universe as a whole, astronomers tell us that the *density* of the stars in the nighttime sky is constant in all directions. By "constant," we mean that if we sample an equal volume of space in any direction, the number of stars in this volume will be approximately equal.

Thus, a better way for us to think about density is as the ratio of the amount of "something" (mass, people, electric charge, etc.) to the size of the spatial region occupied by the "something." This spatial region can be one-, two-, or three-dimensional. In other words, density is nothing more than a measurement of how tightly packed our physical quantity is in a particular spatial region.

Surface Charge Density and Parallel-Plate Capacitors

The first electronics application we will consider is an examination of the surface charge density of a parallel-plate capacitor.

A parallel-plate capacitor is built from two conducting surfaces, each of which has a surface area of A, that are placed close to one another.

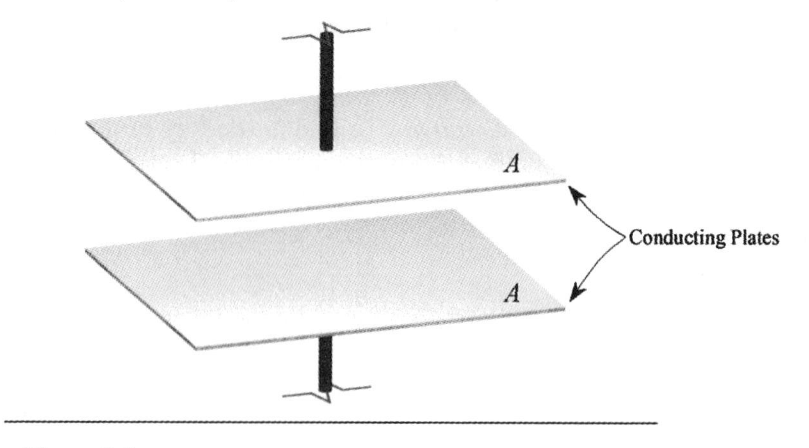

Figure 3.1

When the capacitor is wired to a source voltage, charges begin to accumulate on the plates of the capacitor. Thus, a potential difference, or voltage, appears *across* the capacitor and an electric field appears *between* the plates of the capacitor. The size of both the voltage and the electric field are proportional to the amount of stored charge. As charges are being stored, they are evenly distributed across the plate because of the natural electrostatic repulsion between the negatively charged electrons.

FYI

It should be noted that there are "fringing" effects of the electric field around the edges of the capacitor. For our purposes, it is sufficient to ignore these effects and assume that all electric field lines point straight from the positive plate to the negative plate of the capacitor.

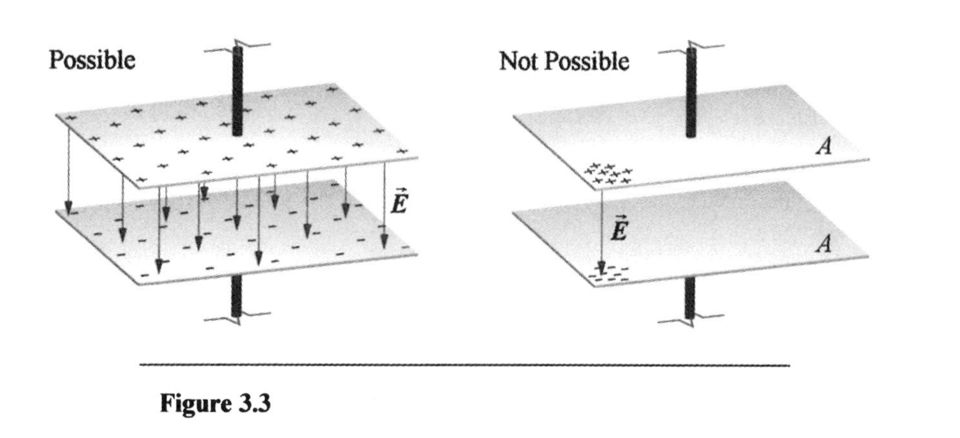

Figure 3.3

Because the charges are uniformly distributed over the entire plate of the capacitor, we can define a quantity that measures how tightly packed the charges are on the plate of the capacitor. This quantity is called *surface charge density* (σ), and is the ratio of the amount of charge stored to the area of the conducting plate:

$$\sigma = \frac{q}{A}$$

Using our new quantity, the electric field inside a parallel-plate capacitor can be expressed as

$$E = \frac{\sigma}{\varepsilon_0} \qquad \textbf{(1)}$$

Electronics Review

Notice that in Equation (1) the right side contains the familiar constant ε_0 that often appears whenever electric fields are present in electronics.

$$\varepsilon_0 = 8.85 \times 10^{-12}\ \frac{\text{F}}{\text{m}}$$

Example 2

Find the electric field between the plates of a capacitor that have accumulated a charge of 5×10^{-10} C and have a cross-sectional area of 0.0001 m^2.

Solution To solve this problem, we must first find the surface charge density on the capacitor,

$$\sigma = \frac{q}{A}$$

$$\sigma = \frac{5 \times 10^{-10}\ \text{C}}{0.0001\ \text{m}^2}$$

$$\sigma = 5 \times 10^{-6}\ \frac{\text{C}}{\text{m}^2}$$

and insert this value into the expression for the electric field:

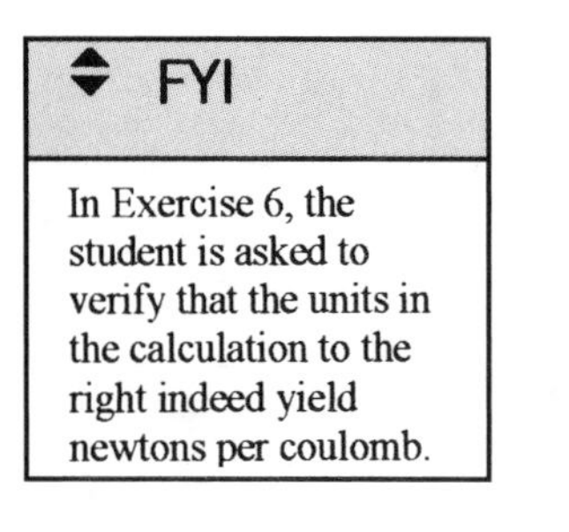

FYI

In Exercise 6, the student is asked to verify that the units in the calculation to the right indeed yield newtons per coulomb.

$$E = \frac{\sigma}{\varepsilon_0}$$

$$E = \frac{5 \times 10^{-6}\ \frac{\text{C}}{\text{m}^2}}{8.85 \times 10^{-12}\ \frac{\text{F}}{\text{m}}}$$

$$E = 5.65 \times 10^{5}\ \frac{\text{N}}{\text{C}}$$

We find that the electric field between the plates of our capacitor is 5.65 x 10^5 newtons per coulomb.

In summary, we see that a broad interpretation of the concept of density, such as this one in which we analyze the amount of charge on a conducting plate, provides a useful tool that can be applied to electronics.

Current Density

Connection

The drift velocity of an electron is discussed in more detail in Application 4 - *Vectors and Electronics.*

To further illustrate the power of the concept of density in electronics, let's apply the concept of density to current flowing through wires of different gauges.

Although the actual path of a conduction electron is not smooth, it is still useful to compare the electric current flowing through a wire to a fluid current Using this analogy, we can represent the amount of electric current that flows through a certain gauge of wire with the same type of diagram that is used for a fluid.

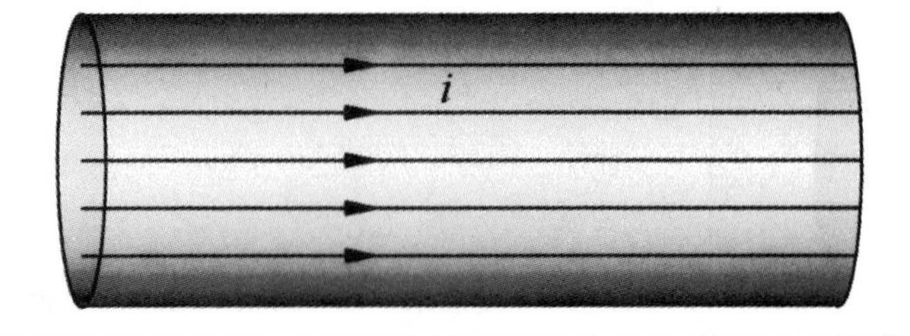

Figure 3.4

Notice that as the cross-sectional area of the wire decreases, the "flow lines" associated with the current become closer to one another.

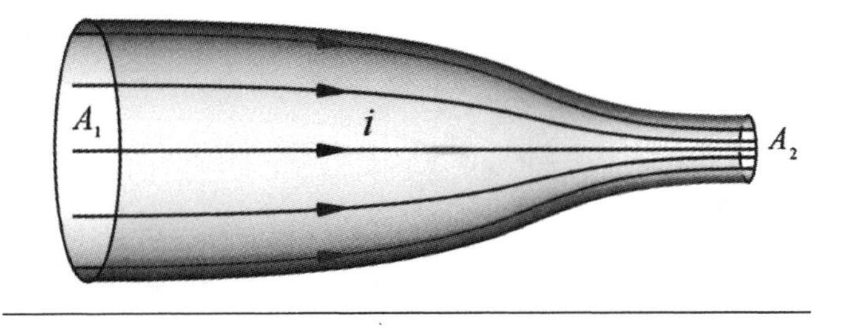

Figure 3.5

Thus, we again find ourselves in the situation where we can define a type of density. In this case, we define a quantity known as current density (J) that is the ratio of the amount of electric current (i) to the cross-sectional area (A) through which it passes

$$J = \frac{i}{A}$$

Example 3

Find the electric current density associated with a current of 5 mA flowing through a wire that has a cross-sectional area of 0.0001 m^2.

Solution To solve this problem we insert the given values into the expression for the current density:

$$J = \frac{i}{A}$$

$$J = \frac{5 \times 10^{-3}\ \text{Amps}}{0.0001\,\text{m}^2}$$

$$J = 50\frac{\text{Amps}}{\text{m}^2}$$

Conclusion

Although we have only discussed two examples of the application of density to electronics, there are many others. As was stated in the introduction to this section of the text, physical principles such as density underlie all of the behavior of electronic devices and circuitry. Knowledge of the *physical* origins of electronic phenomena such as surface charge density and current density provides anyone interested in electronics with a truer understanding of the topic.

Exercises

1. What is the surface charge density of a conducting plate that is 1 cm x 1 cm and has 0.0005 C of charge on its surface?

2. What is the current density in a wire that has a cross-sectional area of 0.01 cm^2 and a current flow of 4 mA?

3. Assume that the area of the wire in Exercise 2 is decreased by half. Calculate the new value of the current density.

4. What value of surface charge density would a parallel-place capacitor need in order to have an electric field of 1 N/C between its plates?

5. If 0.0004 C of charge are evenly spaced inside a solid conducting sphere of radius 0.01 m, find the *volume* charge density of the sphere. Recall that the volume of a sphere is given by

$$V = \frac{4}{3}\pi r^3$$

6. Beginning with the units for the surface charge density, σ, and the permittivity, ε_0, use the equation

$$E = \frac{\sigma}{\varepsilon_0}$$

to show that the units for electric fields can be expressed in newtons per coulomb (N/C).

ELECTRONICS 4

Vectors and Electronics

In this application, we study the difference between scalar and vector quantities in an electronics setting and illustrate the usefulness of vectors in electronics by deriving the expression for the impedance of an RLC-circuit.

Scalars and Vectors – The Drift Velocity of a Conduction Electron

For convenience, we usually draw the current flow through a wire as a simple arrow. However, the actual paths of the conduction electrons that compose the current are far from being straight lines. Because of the repulsion between the conduction electrons and the electrons contained in the wire, rather than being a straight line, the actual path of a conduction electron is actually closer to the one drawn in the figure.

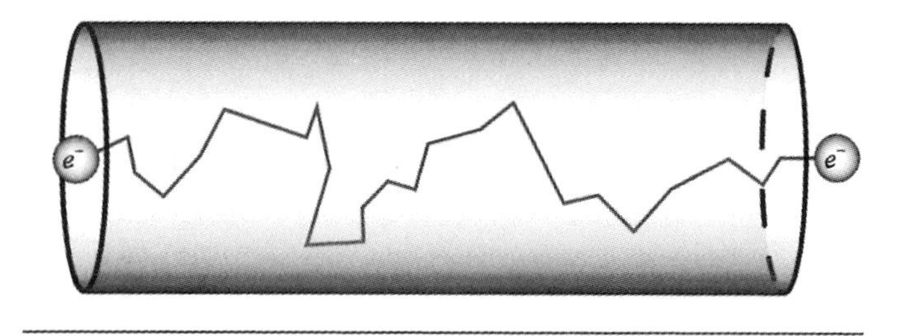

Figure 4.1

We can use the motion of these conduction electrons as a setting to examine the concept of velocity and to analyze the difference between vector and scalar quantities.

Speed and Velocity

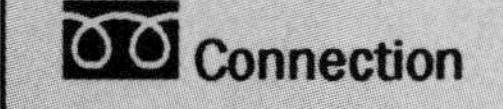

Speed is introduced in Section 2.1 of *College Physics*, and velocity is introduced in Section 2.2.

The authors of *College Physics* distinguish between the physical quantity of velocity, which includes both magnitude and direction, and its scalar counterpart, speed, which includes magnitude alone. Before moving on to our application, it is useful to restate the definitions for velocity and speed:

- *Velocity* is the change in the displacement of an object divided by the elapsed time

- *Speed* is the change in the distance traveled by an object divided by the elapsed time.

Because of the different meanings of these two terms, we can distinguish between the *speed* with which an electron moves in a wire, and the *velocity* of the electron as it moves through a net displacement, $\Delta\vec{x}$, along the wire.

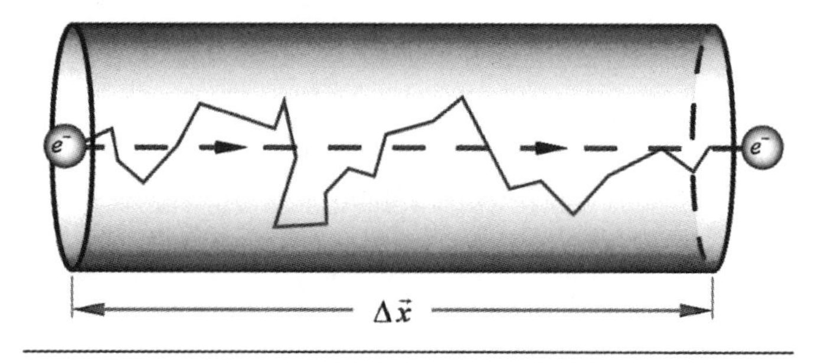

Figure 4.2

The velocity of the electron in the wire, which can be expressed as the net linear displacement of the electron divided by the time that it takes to execute this motion,

$$\vec{v}_{\mathrm{D}} = \frac{\Delta\vec{x}}{\Delta t}$$

is known as the *drift velocity* of the electron. This velocity is an illustration of the difference between vector and scalar quantities. Although a conduction electron can cover a great deal of *distance* as it executes its motion, it is only the net *displacement* that enters into the drift velocity calculation. Therefore, when analyzing the flow of current in a wire and the amount of time required for the current to move from one point in a circuit to another, it is the smaller *drift velocity* that must be used rather than the actual high *speed* of the conduction electrons.

▸ Vectors in Electronics – The Impedance Triangle

To demonstrate further the applicability of vectors in electronics, we will use the setting of an RLC-circuit, familiar from ac electronics.

Recall that the alternating voltage, V_{AC}, the current flow, I, and the impedance in the circuit, Z, are related by the equation

Electronics Review

In the impedance equation, X_C is the capacitive reactance and X_L is the inductive reactance.

$$V_{AC} = IZ$$

and the impedance in the circuit is found using the equation

$$Z = \sqrt{R^2 + (X_L - X_C)^2}$$

Connection

Methods of vector addition are introduced in Section 3.2 of *College Physics.*

Our purpose in this section is to use the methods of vector addition introduced in *College Physics* to derive this familiar impedance equation.

In an RLC-circuit like that illustrated in Figure 4.3, the voltage across each circuit element alternates with the same frequency as the source voltage.

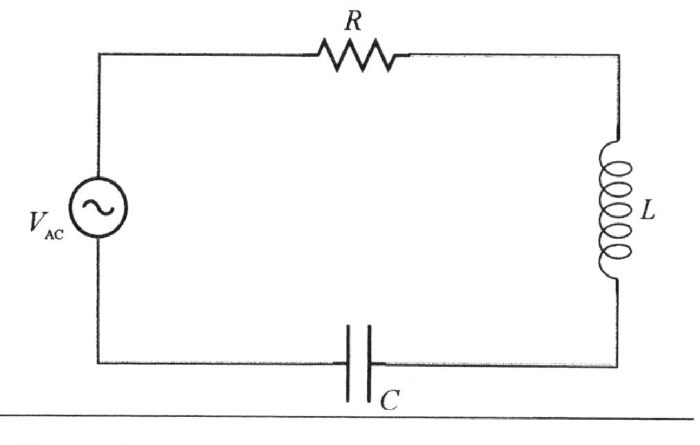

Figure 4.3

Although each element's voltage oscillates with the same frequency, the voltages are not in phase with one another. The voltage across the capacitor, V_C, lags the voltage across the resistor, V_R, by 90°, and the voltage across the inductor leads the voltage across the resistor, V_C, by 90°.

Instead of making a complex graph to summarize the relationship among these three sine waves, a *phasor diagram* is often used to illustrate these relationships. Phasors obey the same mathematical rules as the vectors introduced in *College Physics*. The only difference between these two mathematical tools is that phasors rotate around the graph with a frequency determined by the source voltage instead of remaining fixed, as do the vectors. The length of each phasor, denoted with a capital letter, is determined by the amplitude of the voltage across the circuit element in the same way as the length of a vector is determined by the magnitude of the physical quantity that the vector represents. To find the value of the voltage at a particular instant, denoted by a lower-case letter, we find the projection of the phasor along the vertical axis:

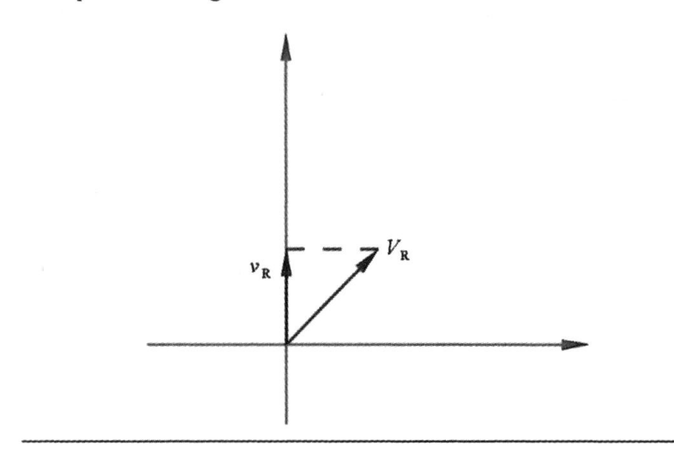

Figure 4.4

A typical phasor diagram expressing the lead/lag relationships between the voltages in an RLC-circuit is given in Figure 4.5:

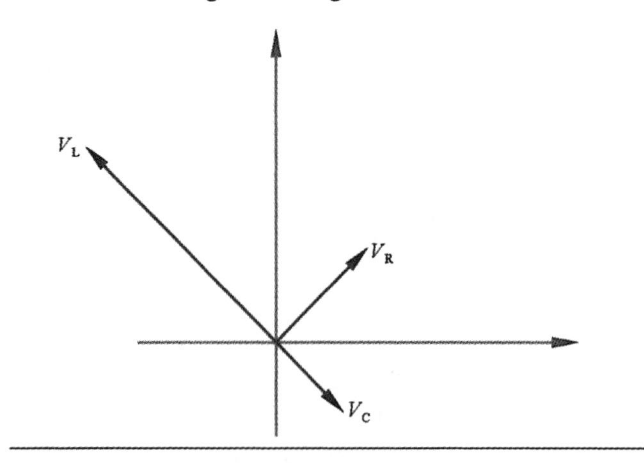

Figure 4.5

Because the current and the voltage across the resistor are in phase with one another, it is common in the electronics field to use V_R as the reference and orient the phasors in Figure 4.5 along the x- and y-axes:

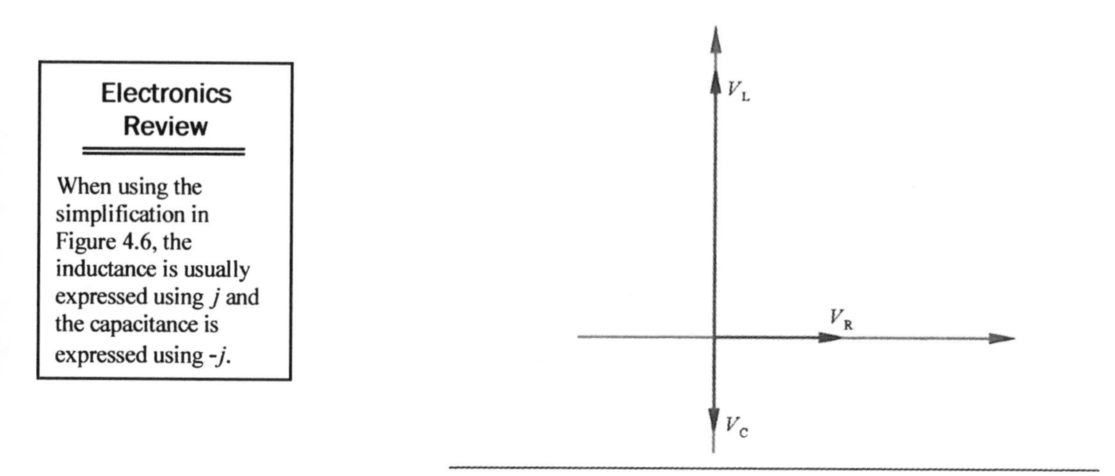

Figure 4.6

Although we could use either Figure 4.5 or Figure 4.6 as a starting point for our derivation of the impedance expression, the simplification in Figure 4.6 hides the fact that the phasors rotate around the diagram with a frequency determined by the source voltage. Because of this, we will use Figure 4.5 to begin our derivation of the impedance expression.

Connection

The methods of vector addition are discussed in 3.2 of *College Physics.*

Beginning with Figure 4.5, notice that since V_L and V_C represent the same type of physical quantity, voltage, we can subtract these phasors from one another.

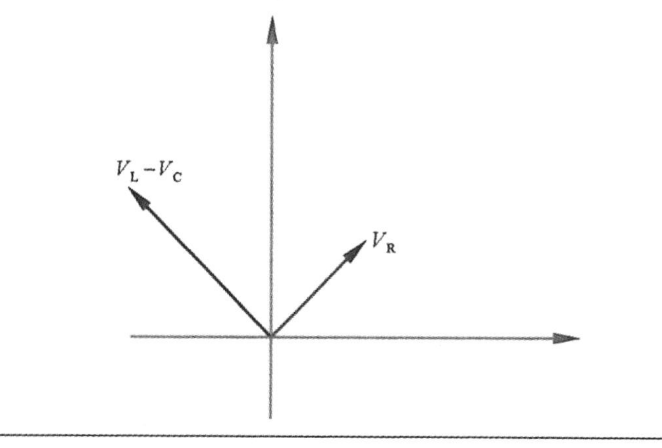

Figure 4.7

From Kirchhoff's Voltage Law, we know that the algebraic sum of the voltages across the elements in a circuit must be equal to the source voltage. Expressed in the language of vectors and phasors, this means that at any instant the algebraic result of the two phasors V_L-V_C and V_R must be equal in magnitude to the source voltage, V_{AC}:

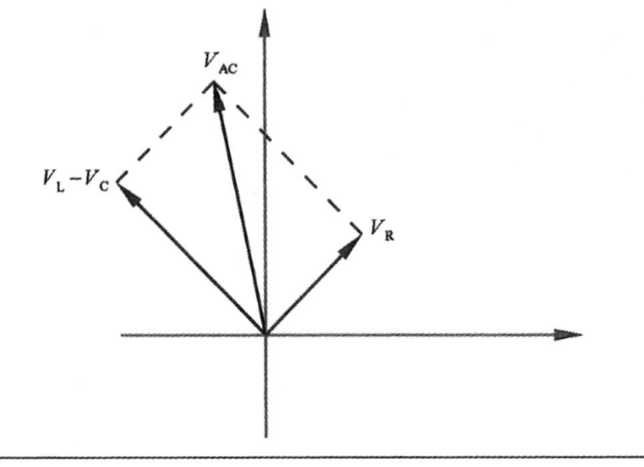

Figure 4.8

To find the source voltage, we can form a right triangle that will allow us to use the Pythagorean Theorem to find an appropriate expression:

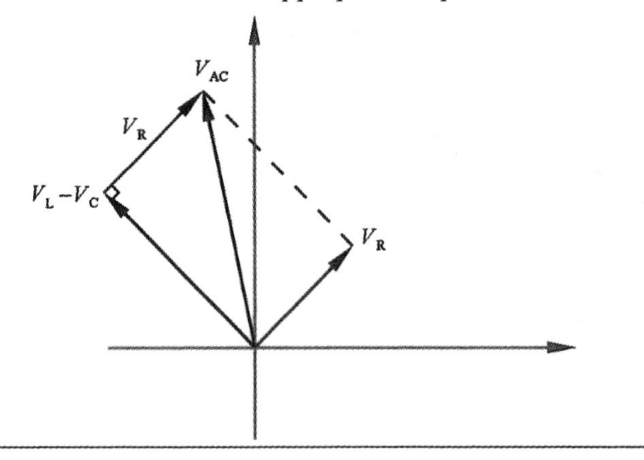

Figure 4.9

Substituting the voltage expressions into the Pythagorean Theorem yields

$$V_{AC}^2 = V_R^2 + (V_L - V_C)^2 \quad \textbf{(1)}$$

Recalling that each of the voltage amplitudes is related to the current amplitude, I, by the following equations:

$$V_{\mathrm{R}} = IR$$
$$V_{\mathrm{C}} = IX_{\mathrm{C}}$$
$$V_{\mathrm{L}} = IX_{\mathrm{L}}$$

Equation (1) becomes

$$V_{\mathrm{AC}}^2 = (IR)^2 + (IX_{\mathrm{L}} - IX_{\mathrm{C}})^2$$

This equation can now be rearranged in the following way:

$$V_{\mathrm{AC}}^2 = (IR)^2 + (IX_{\mathrm{L}} - IX_{\mathrm{C}})^2$$
$$V_{\mathrm{AC}}^2 = I^2R^2 + I^2(X_{\mathrm{L}} - X_{\mathrm{C}})^2$$
$$V_{\mathrm{AC}}^2 = I^2\left[R^2 + (X_{\mathrm{L}} - X_{\mathrm{C}})^2\right]$$
$$V_{\mathrm{AC}} = \sqrt{I^2\left[R^2 + (X_{\mathrm{L}} - X_{C})^2\right]}$$

to yield

$$V_{\mathrm{AC}} = I\sqrt{R^2 + (X_{\mathrm{L}} - X_{\mathrm{C}})^2} \qquad \textbf{(2)}$$

If we now compare Equation (2) with the equation relating voltage, current, and impedance that opened this section,

$$V_{\mathrm{AC}} = IZ$$

we see that the impedance, Z, can be expressed as

$$Z = \sqrt{R^2 + (X_{\mathrm{L}} - X_{\mathrm{C}})^2}$$

▸ Conclusion

Based on our examination of the drift velocity of a conduction electron and our derivation of the equation for the impedance of an RLC-circuit, we see that vectors are a useful tool that can help us develop a deeper understanding of electronics. An understanding of vectors gives insights into the physical quantities in electronics, such as drift velocity, as well as the relationships between quantities, such as the voltages across several elements in the same circuit.

Exercises

1. Find the drift velocity of an electron if it moves along a wire a net displacement of 1 m in 1 μs.

2. If the drift velocity of a certain conduction electron is 4.5 x 10^6 m/s, through what length of wire will the electron move in 5 seconds?

3. Discuss how the concept of drift velocity is related to minority charge carrier current in a pn-junction diode.

4. Discuss how the concept of drift velocity is related to the avalanche effect in a pn-junction diode.

5. Since an electron in an alternating-current circuit moves approximately ten atoms in either direction before changing directions, how does the concept of drift velocity apply to this type of circuit?

6. Find the impedance of a series RLC-circuit containing a 5 KΩ resistor, a 4 pF capacitor, and a 6 μH inductor if the source voltage is driving the circuit at 110 Hz.

7. Using Figure 4.9 and the fact that the current phasor is in phase with the voltage across the resistor, show that the phase angle for the circuit, φ, can be found using

$$\tan\varphi = \frac{X_{\mathrm{L}} - X_{\mathrm{C}}}{R}$$

ELECTRONICS 5

The Force on an Electron in an Electric Field

In this short application, we use the setting of an electron in a current-carrying wire to illustrate the concepts of force, vectors, and scalars.

The Force on a Charge in an Electric Field

When the switch is closed in an electric circuit, an electric field is produced inside the wire:

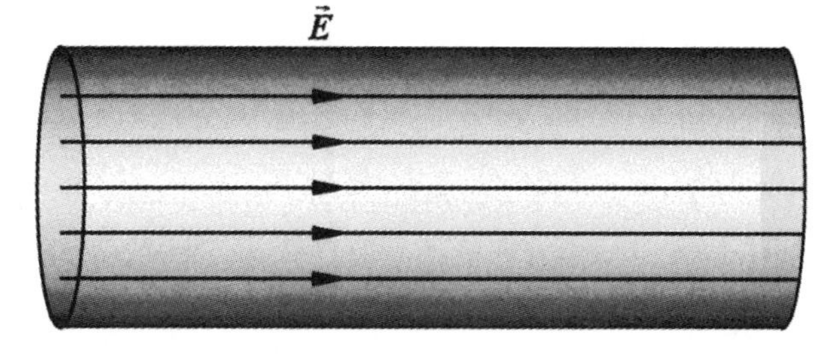

Figure 5.1

Just as the polarity of the applied voltage can be positive or negative, the direction of the electric field inside the wire can be directed to the right or to the left:

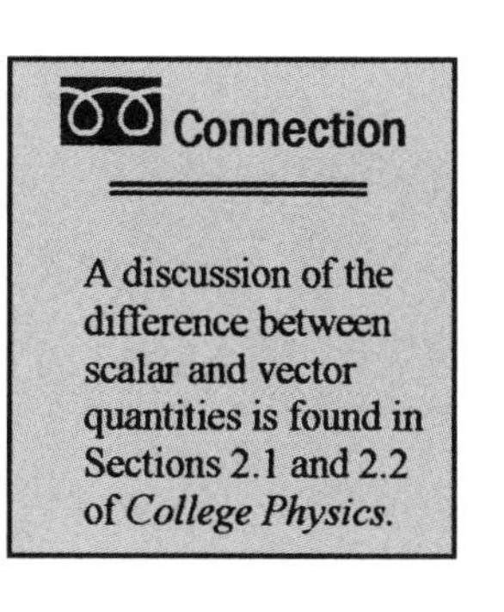

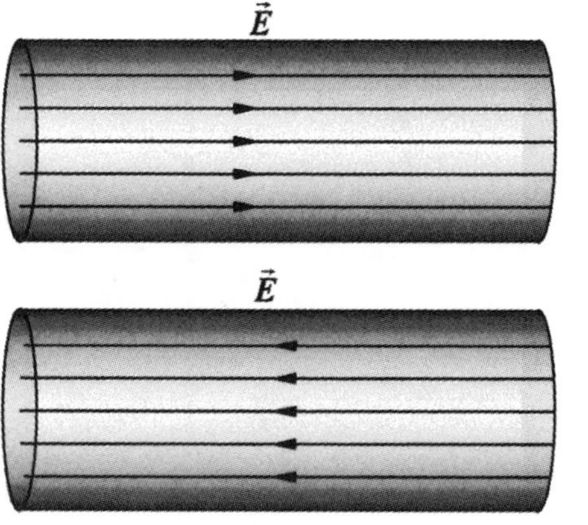

Figure 5.2

Connection

Force is introduced in Chapter 4 of *College Physics.*

When an electric charge, q, is exposed to an electric field, the charge experiences a force. This force is found by multiplying the charge by the electric field:

$$\vec{F} = q\vec{E} \qquad \textbf{(1)}$$

This equation states that the size of the force experienced by the electric charge is proportional to both the size of the electric charge and the size of the electric field in which the charge is placed.

Because the electric field has a direction associated with it, the force that the charge experiences will also have a direction. Equation (1) is based on the force that a *positive* charge, q, would experience in the electric field:

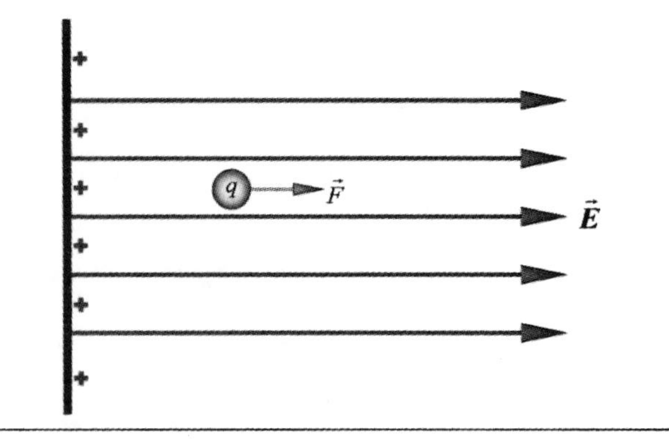

Figure 5.3

If we now focus on the force that an *electron* inside a current-carrying wire would experience, we see that the force is opposite to the direction of the applied electric field because of the negative sign attached to the charge on the electron:

$$\vec{F} = (-1.6\times10^{-19}\,\text{C})\vec{E}$$

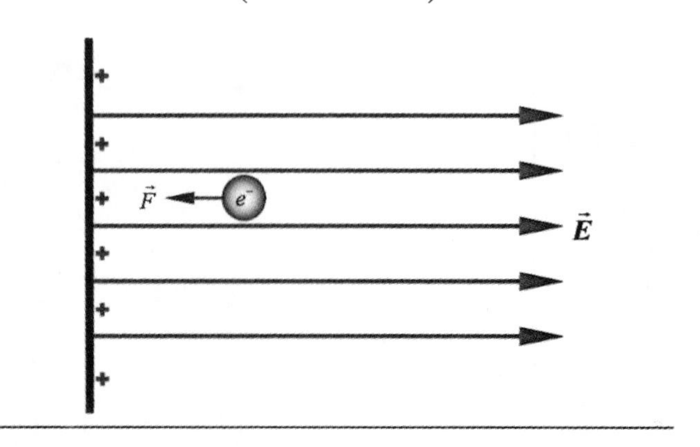

Figure 5.4

Because the physical quantity of charge does not have an associated direction, it is referred to as a *scalar*. Conversely, because we must know the direction of the electric field and the force to completely define each of these physical quantities, they are known as *vector* quantities.

Example 1

Find the magnitude and direction of the force on an electron that is placed in an electric field that is pointing to the right and has a magnitude of 100 N/C.

Connection

The relationship between force and acceleration is expressed by Newton's Second Law

$$\sum \vec{F} = m\vec{a}$$

discussed in Section 4.3 of *College Physics.*

Solution We solve the problem by inserting our known values into Equation (1)

$$\vec{F} = (-1.6\times10^{-19}\ \text{C})(100\ \text{N/C})$$
$$\vec{F} = -1.6\times10^{-17}\ \text{N}$$

Notice that the force on the electron, and hence the direction of its acceleration, is opposite to the direction of the applied electric field. Thus, the electron experiences a force to the left due to the electric field.

Applications of the Forces on Electric Charges

The fact that an electric field exerts a force on an electron is of great importance in many electronics situations. In this section, we will use the information we learned about the force on electric charges in electric fields and apply it in two familiar electronics settings: the depletion layer inside a pn-junction diode, and the dielectric materials inserted between the plates of a capacitor.

The Size of the Depletion Layer Inside a PN-Junction Diode

The *depletion layer* inside a pn-junction diode is an area located in the central region of the diode that has been depleted of mobile charge carriers. In the case of a gallium-arsenide junction, this region arises as electrons from the n-side of the diode fill holes in the p-side:

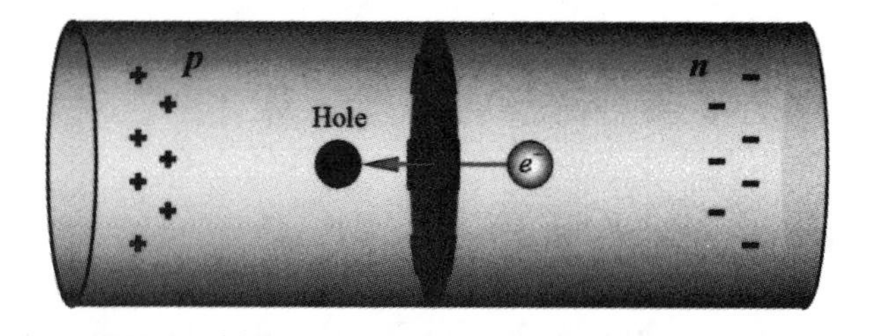

Figure 5.5

This movement of electrons from the n-side to the p-side creates a region in the material that is bounded on either side by positive and negative ions. This region is the *depletion layer* familiar from solid-state electronics. As the depletion layer grows, a point is reached at which no more electrons can migrate from the n-side to the p-side. Our newly acquired understanding of the force on electric charges in electric fields will allow us to give a physical explanation for why the migration of charges cannot continue until all of the arsenic atoms have donated an electron.

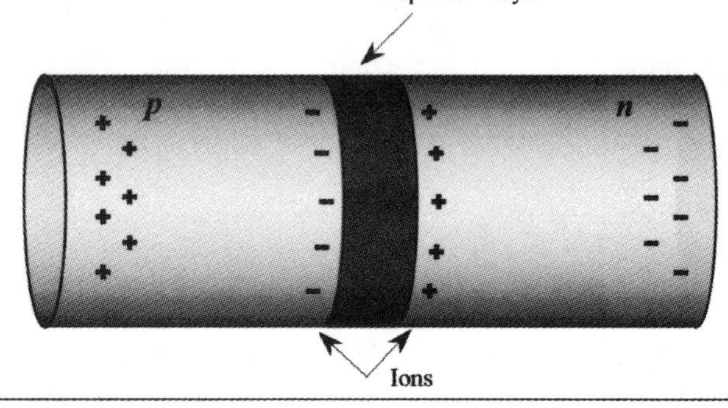

Figure 5.6

Because ions are created in pairs at the interface between the p-type and n-type semiconductors, the depletion layer must be bounded on both sides by the same number of ions. Thus, the depletion layer can be viewed as a very small capacitor. In the same way that a charged capacitor has charges of q and $-q$ on its plates, the depletion layer has equal and opposite charges on either side.

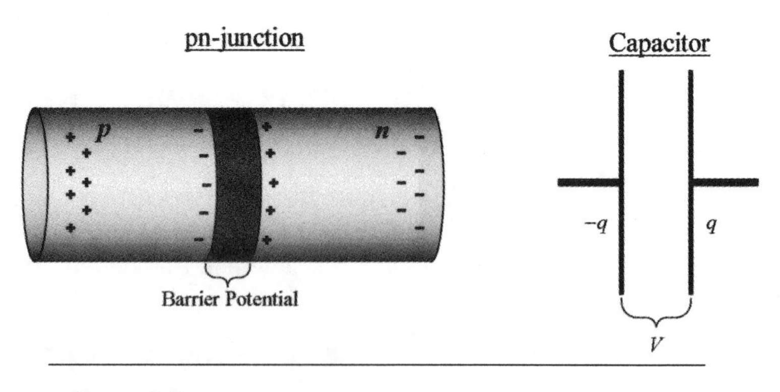

Figure 5.7

Continuing the analogy, the depletion layer also has a potential difference (voltage) across it, just as a charged capacitor has a potential difference across its plates. This potential across the depletion layer is known as the *barrier potential.* More important for our example, however, is the fact that just as a charged capacitor has an electric field between its plates, an electric field appears inside the depletion layer. This electric field continues to grow as the depletion layer is created. Eventually, the electric field exerts so great of a force on the electrons attempting to migrate from the n-side that the junction reaches a state of equilibrium and no more ion pairs can be formed. The depletion layer has then reached its maximum size and acts as an insulator inside the diode. Thus, we see again that it is the force exerted on the electrons by an electric field that plays a pivotal role in determining the behavior of an electronic circuit element.

Dielectric Materials Inside Capacitors

To increase the capacitance of a capacitor, dielectric materials are often inserted between its plates. Certain dielectric materials have a large impact on the capacitance, while others have a small one. In this example, we will use the concept of the force on electric charges in electric fields to explain one way in which dielectric materials increase capacitance.

An *electric dipole* is composed of two equal and opposite charges, q and $-q$, that are separated by a distance of d:

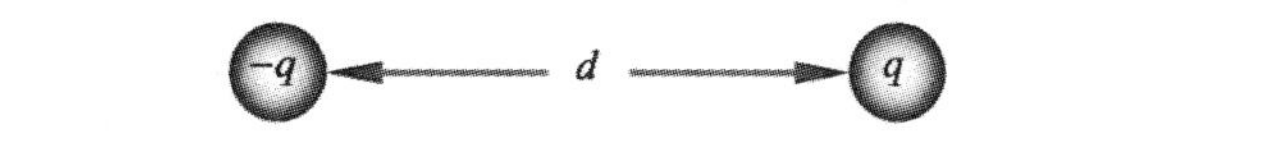

Figure 5.8

If a dipole is placed in an electric field, each charge will experience a force because of the field. Because the charges are equal and opposite, there will be a net force of zero on the dipole:

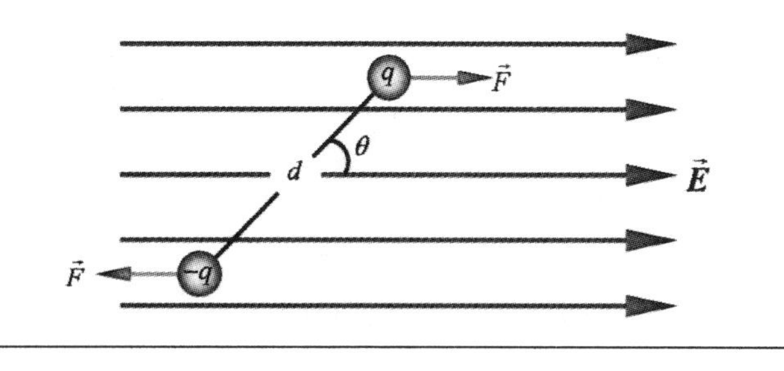

Figure 5.9

However, there will be a *torque* on the dipole as it tries to align itself with the electric field. This situation repeats itself in the case of dielectric materials inserted between the plates of a capacitor. As the capacitor charges, an electric field appears between the plates. This electric field exerts forces on the individual charges that comprise the molecules of the dielectric material.

Although there may not be a net force on the molecules, the distribution of the electric charges that make up the molecules of a dielectric material will produce a net torque that tends to cause the molecules to rotate. Thus, as the capacitor charges, millions of molecules rotate to align themselves with the field. If this field decreases, or disappears, the molecules will snap back to their original orientations in the solid because of the electrostatic bonds in the material. We can therefore think about the torqued molecules as being millions of little springs inside the material. Each "spring" stores potential energy as it is torqued in the field. As the capacitor discharges, the field becomes smaller and the molecules are allowed to return to their original configuration in the solid. In other words, we are able to extract the potential energy from our "springs."

We see then that by placing a dielectric material between the plates of the capacitor, the amount of potential energy that can be stored in the capacitor increases and therefore the capacitance of the capacitor increases.

Conclusion

An understanding of the force on electric charges caused by electric fields is important in our study of physics and electronics. Not only does it help us to understand the motion of electrons inside current-carrying wires, but also to gain a deeper understanding of many other electronic situations, such as the depletion layer inside pn-junction diodes and the ability of a capacitor to store potential energy in a circuit.

Exercises

1. Calculate the magnitude of the force on an electron if it is in an electric field of 100 N/C.

2. The charge on a proton is identical to the charge on the electron except it is positive instead of negative:

$$q_P = 1.6 \times 10^{-19} C$$

   What effect, if any, would this difference in sign have on the force that the proton would experience if it is placed in the same electric field as the electron in Exercise 1?

3. What size of electric field would be required to cause a force of 2 N on an electron?

4. An alpha particle is the nucleus of a Helium - 4 atom. This particle contains two protons and two neutrons (q_N = 0). If an alpha particle is inserted into an electric field of magnitude 200 N/C, find the magnitude of the force on the alpha particle.

5. The vast majority of atoms in our world are electrically neutral since each atom contains as many electrons surrounding the nucleus as protons contained in the nucleus. What does this say about the net force on an average atom if we place it in an electric field?

ELECTRONICS 6

The Force Between Two Current-Carrying Wires

In this application, the electronics setting of two parallel wires is used to illustrate the difference between attractive and repulsive forces.

As is well known from basic electronics, the motion of the conduction electrons inside a wire produces a magnetic field around the wire.

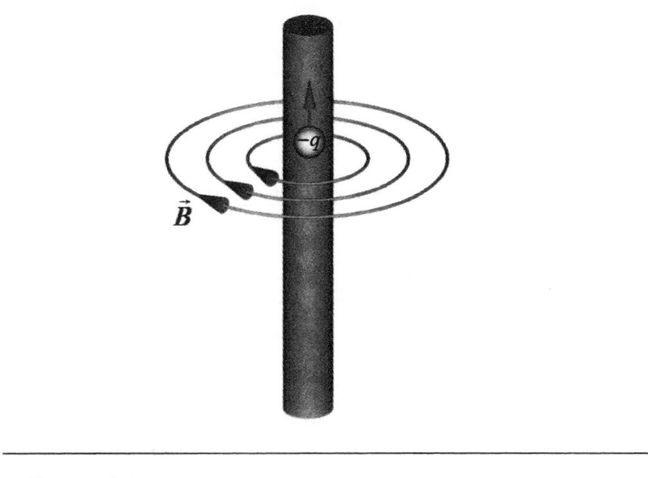

Figure 6.1

What is not quite as familiar to many electronics students is the reverse relationship that a magnetic field exerts a force on a moving charge and thus on a wire carrying current. This force, given by

$$F = qvB\sin\theta \qquad \textbf{(1)}$$

is dependent on the size of the magnetic field, B; the size of the charge moving through the field, q; and the orientation of the velocity, v, of the charge with respect to the magnetic field:

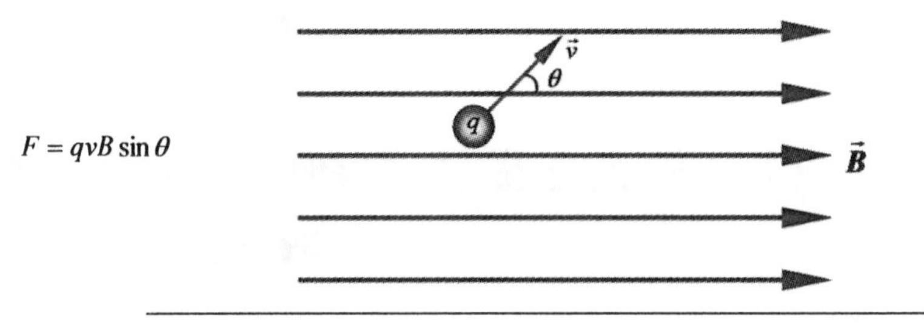

Figure 6.2

> **Math Review**
>
> Remember that
> $\sin 90° = 1$
> $\sin 0° = 0$

Thus, a charge moving at right angles to the field lines experiences the maximum force while a charge moving *along* a field line experiences zero force:

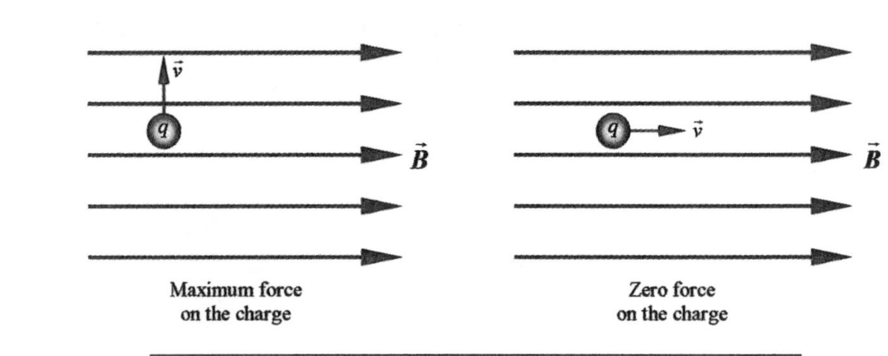

Figure 6.3

Consequently, if we place a current-carrying wire in a magnetic field, each of the conduction electrons in the current will experience a force because of the field. If we put *two* such wires close to one another, the moving charges in one wire will cause a magnetic field that affects the moving charges in the other wire. Therefore, each wire will experience a force due to the presence of the other wire.

Because of the immense number of individual charges in even a small current, it is simpler for us to rewrite Equation (1) in terms of the total current flow, i, and the length of the wire, L, instead of the individual charges, q, with their velocities. If we assume our wires to be straight, this rewrite becomes

$$F = iLB \sin \theta \qquad \textbf{(2)}$$

To be confident that Equation (2) actually takes into account the charges and their velocities, we need only compare the units on the product iL with those on the product qv.

Connection

Units and unit analysis are discussed in Electronics Application 2 - *Unit Analysis and Conversions in Electronics,* as well as in Sections 1.2, 1.3, and 1.4 of *College Physics.*

Beginning with the SI standard units for the product iL,

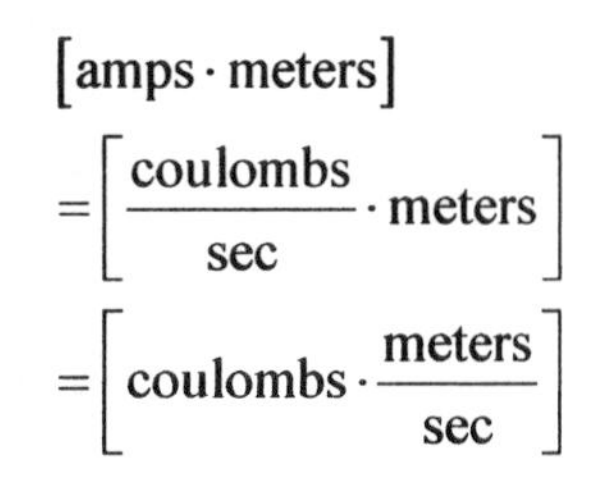

$$\left[\text{amps}\cdot\text{meters}\right]$$
$$=\left[\frac{\text{coulombs}}{\text{sec}}\cdot\text{meters}\right]$$
$$=\left[\text{coulombs}\cdot\frac{\text{meters}}{\text{sec}}\right]$$

Since coulombs is the unit of charge, q, and meters per second is the unit for velocity, v, we see that the units for iL are indeed the same as those for qv.

If the two wires are now placed parallel to one another, the motion of the current in each wire, and hence the electric charges, makes a 90° angle with the magnetic field produced by the other wire:

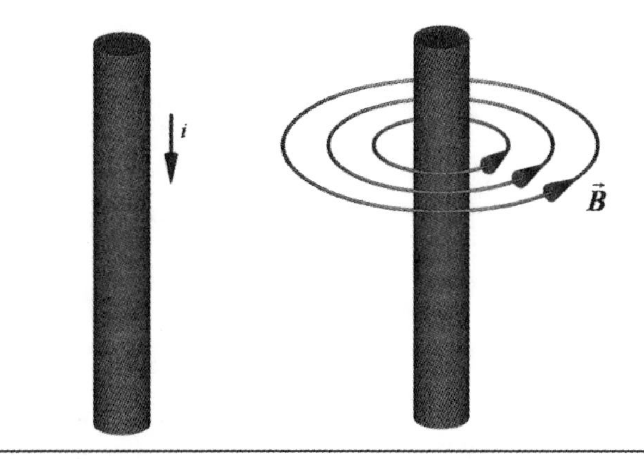

Figure 6.4

Because this situation requires us to take the sine of 90° and sin90° = 1, Equation (2) becomes

$$F = iLB \qquad \textbf{(3)}$$

Notice that because current can flow one of two directions through a wire, the force exerted on our wire can either be to the right or to the left:

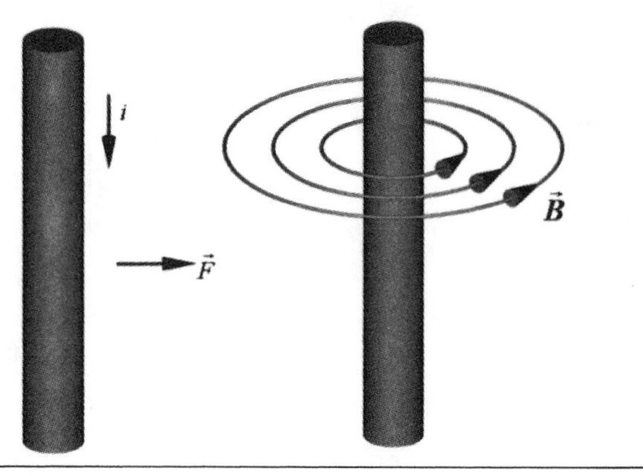

Figure 6.5

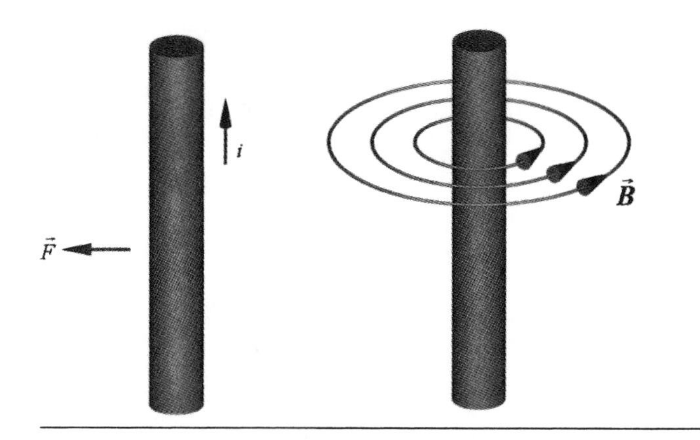

Figure 6.6

Thus, if the current is flowing in the same direction through two parallel wires in a circuit, an attractive force will be generated between the wires:

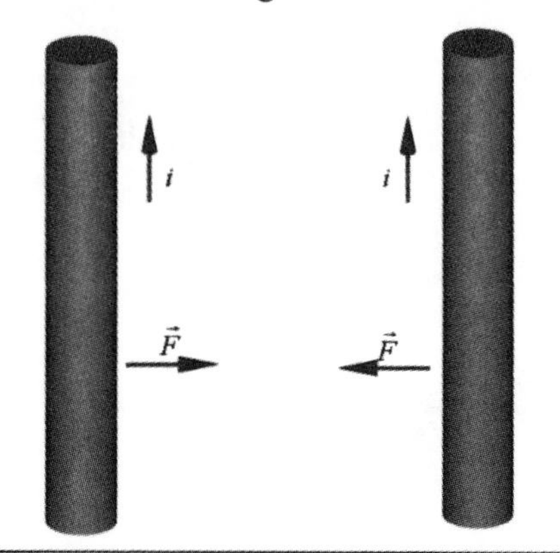

Figure 6.7

Conversely, if the current flows are opposite to one another, the force between the wires will be repulsive:

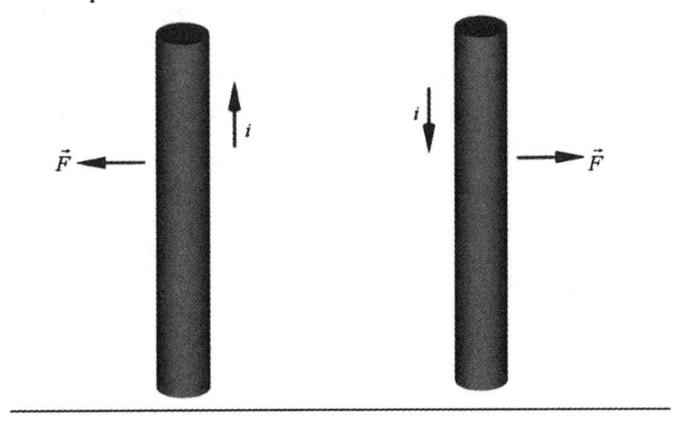

Figure 6.8

Example 1

Two wires in a circuit are placed parallel to one another with their currents running in the same direction. If a current of 5 mA flows through the left wire, and it is exposed to a magnetic field of 6×10^{-7} T because of current flow in the right wire, find the direction and magnitude of the force on the left wire.

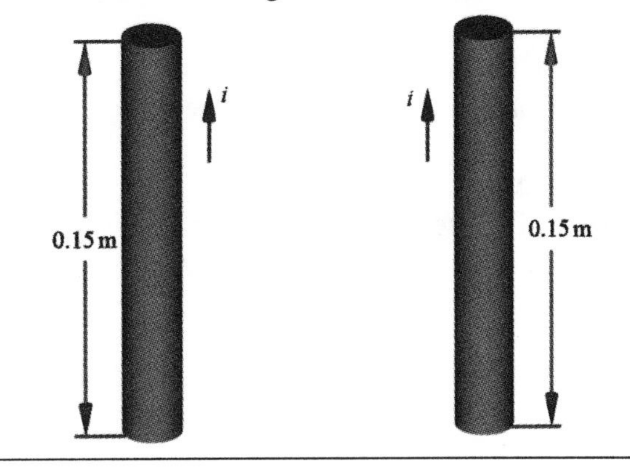

Figure 6.9

Solution The answer is found by inserting the given values into Equation (3):

$$F = iLB$$

$$F = (0.005 \text{ A})(0.15 \text{ m})(6 \times 10^{-7} \text{ T})$$

$$F = 4.5 \times 10^{-10} \text{ N}$$

Since the currents are flowing in the same direction, the left wire will be pulled to the right.

Conclusion

In closing, let us draw attention to the size of the force we calculated in this example. For many circuits this force is negligible. However, when dealing with large current flows or large external magnetic fields is it necessary to consider this force between the wires.

Exercises

1. Find the force on a 1 C charge if it is moving at 5 m/s in a magnetic field of 1 T when each of the following conditions apply:

   a) the charge is moving at right angles to the magnetic field lines
   b) the charge is moving along a magnetic field line
   c) the velocity of the charge is making an angle of 30° with respect to the magnetic field line

2. What, if any, effect would there be on the force if a charge of -1 C were used in the problem statement given in Exercise 1?

3. Find the magnitude of the force on a 1m length of wire if it has 5 mA of current flowing through it and is exposed to a magnetic field of 0.0001 T.

4. In old televisions, the picture could be greatly distorted by placing a magnet on the screen of the set. Using the topic of this application, explain why a magnet would cause this type of distortion.

5. Three wires with 6 mA of current are placed parallel to one another and are separated by 0.01 m. Find the force on the middle wire for each of the current flow configurations.

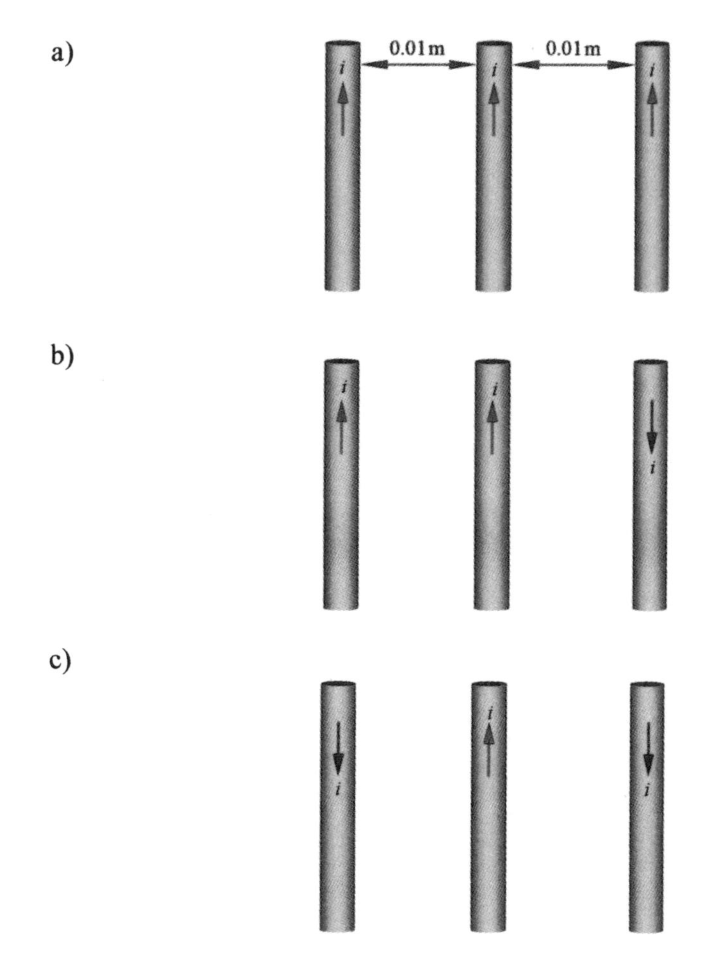

6. ***Challenge Problem***

The magnetic field of a straight current-carrying wire can be found using the equation

$$B = \frac{\mu_0 i}{2\pi r}$$

where r is the distance at which we are measuring the magnetic field from the wire, μ_0 is the permeability of free space, $4\pi \times 10^{-7}$ Tm/A, and i is the current flow through the wire. Using this equation and Equation (3) from this application, find the magnitude of the force on a wire carrying 4 mA of current if the wire is placed 0.01 m away from a parallel wire carrying 5 mA of current.

ELECTRONICS 7

A Comparison of the Gravitational and Electromagnetic Forces

In this application, we contrast the relative strengths of the gravitational and electromagnetic forces at various size scales.

The two forces of gravitation and electromagnetism both operate on size scales ranging from the subatomic to the galactic; however, which of the two forces dominates in a given situation depends upon the size of the system being analyzed. In this application, we contrast these two forces to illustrate why one of them can often be ignored depending upon the size scale of the problem. We begin with a brief review of the force laws associated with each of the forces and then apply the laws to two examples at different size scales.

Newton's Law of Gravitation

> **Connection**
>
> A discussion of Newton's Law of Gravitation is found in Section 7.5 of *College Physics.*

In 1687, Sir Isaac Newton published an equation for calculating the amount of attractive gravitational force between two masses. This equation states that the amount of gravitational force (F) becomes greater as the magnitude of the two masses increases. Conversely, the amount of force decreases as the separation distance (r) between the masses increases.

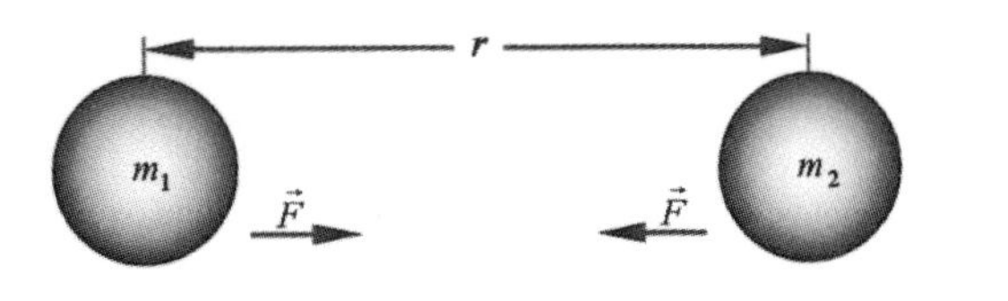

Figure 7.1

FYI

The experiment that Henry Cavendish conducted to measure the gravitational force between two masses, and thus find a numerical value for the gravitational constant, was not performed until over 100 years after Newton published his force law.

Newton's Law of Gravitation can be stated in equation form as

$$F = \frac{Gm_1m_2}{r^2}$$

where

- F is the amount of gravitational force between the masses, expressed in newtons
- m_1 and m_2 are the masses, expressed in kilograms
- r is the separation distance between the masses, expressed in meters
- G is a constant of proportionality known as the Universal Gravitational Constant. This constant, first measured by Henry Cavendish, is given by

$$G = 6.67 \times 10^{-11} \frac{\text{Nm}^2}{\text{Kg}^2}$$

Coulomb's Law

While electronics students learn early in their studies that like electric charges repel one another and opposite charges attract one another, they often do not calculate the actual *size* of this repulsion or attraction.

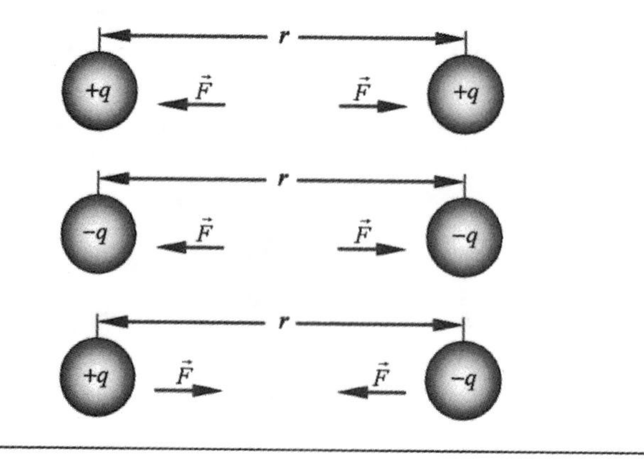

Figure 7.2

FYI

Although the *electromagnetic* force of nature is at the heart of Coulomb's Law, we refer to the force as *electrostatic* when discussing charges that are not moving relative to one another.

Coulomb's Law, created by August Coulomb, can be used to find an actual numerical value for the amount of electrostatic force between two charges. Coulomb's Law is given by

$$F = \frac{kq_1q_2}{r^2}$$

where

- F is the amount of electrostatic repulsion/attraction between the two charges, expressed in newtons
- q_1 and q_2 are the magnitudes of the two electric charges, expressed in coulombs
- r is the separation distance between the charges, expressed in meters
- k is a constant of proportionality known as the Universal Electrostatic Constant. This constant has a value of

$$k = 8.99 \times 10^9 \, \frac{\text{Nm}^2}{\text{C}^2}$$

Notice that Coulomb's Law is amazingly similar to Newton's Law of Gravitation:

Newton's Law of Gravitation:	$F = \frac{Gm_1m_2}{r^2}$
Coulomb's Law:	$F = \frac{kq_1q_2}{r^2}$

Thus, while the gravitational force between the *two masses* is proportional to the size of the masses and inversely proportional to the square of the separation distance between them, the electrostatic force is proportional to the magnitude of the *two electric charges* and inversely proportional to the square of the separation between the charges.

Armed with this information, we can now apply these two equations to two different examples at vastly different size scales. As we will see, in the subatomic world the electromagnetic force is much larger and more important than the gravitational force. In contrast, when we reach the size scale of planets and stars, the electromagnetic force becomes negligible and the gravitational force becomes dominant.

Comparing the Gravitational and Electrostatic Forces

Example 1

In this example, we will use Newton's Law of Gravitation to calculate the amount of gravitational attraction between two conduction electrons in a sample of wire. After generating a result, we will work with the same two electrons and calculate the amount of electrostatic force between them using Coulomb's Law.

Find the electrostatic and gravitational forces between the following pair of electrons:

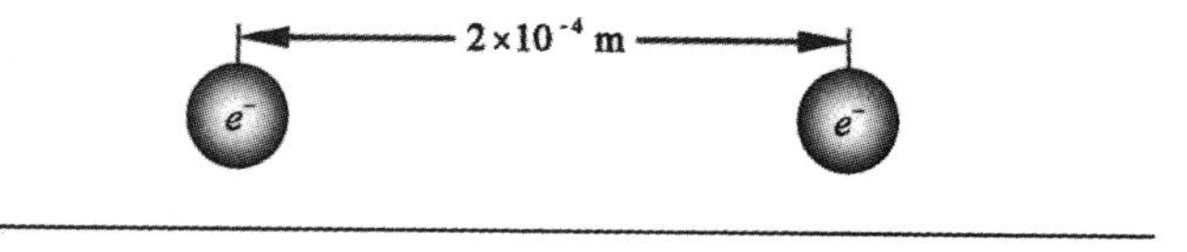

Figure 7.3

Solution First, we must find the magnitude of the electrostatic repulsion between the two electrons.

Beginning with Coulomb's Law

$$F = \frac{kq_1q_2}{r^2}$$

we insert the values of the charge, separation distance, and electrostatic constant to generate the electrostatic force:

> **FYI**
>
> Note that in the calculation to the right q_1 and q_2 have the same value since all electrons possess the same amount of electric charge.

$$F = \frac{kq_1q_2}{r^2}$$

$$F = \frac{(8.99\times10^{9}\ \text{Nm}^2/\text{C}^2)(-1.6\times10^{-19}\ \text{C})(-1.6\times10^{-19}\ \text{C})}{(2\times10^{-4}\ \text{m})^2}$$

$$F = 5.8\times10^{-21}\ \text{N}$$

Notice that this force is repulsive since both electrons possess negative charge.

To find the amount of gravitational attraction between the electrons we use Newton's Law of Gravitation:

> **FYI**
>
> Note that in the calculation to the right both m_1 and m_2 have the same value since all electrons have the same mass.

$$F = \frac{Gm_1m_2}{r^2}$$

$$F = \frac{\left(6.67\times10^{-11}\ \text{Nm}^2/\text{Kg}^2\right)\left(9.11\times10^{-31}\ \text{Kg}\right)\left(9.11\times10^{-31}\ \text{Kg}\right)}{\left(2\times10^{-4}\ \text{m}\right)^2}$$

$$F = 1.38\times10^{-63}\ \text{N}$$

If we now compare the electrostatic force between the two electrons to the amount of gravitational force, we see that they differ by 84 orders of magnitude! This example demonstrates that when doing physical calculations in the world of the subatomic, as we do in electronics, it is nearly always safe to ignore the force of gravitation between the electrons.

Example 2

In the previous example we saw that in the microscopic world the force of gravity is negligible when compared to the electromagnetic force. In this example, we will compare these two forces in a much larger setting.

Contrast the electrostatic and gravitational forces between the earth and the moon.

> **FYI**
>
> Note that the earth-moon distance is a *mean* value since the moon's orbit is not perfectly circular.

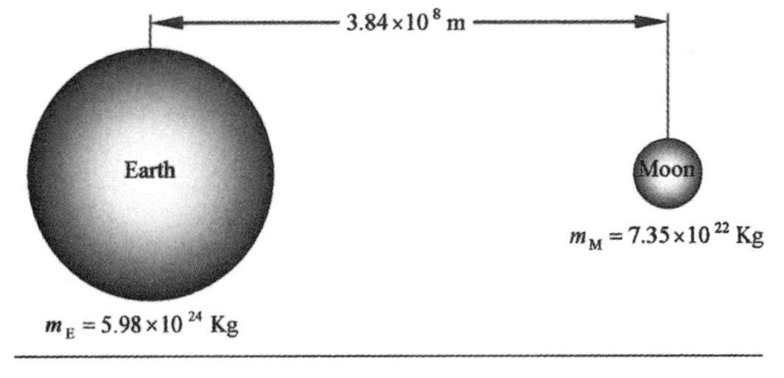

Figure 7.4

Solution The gravitational force between the earth and the moon is found by using Newton's Law of Gravitation and inserting our known values:

$$F = \frac{Gm_1m_2}{r^2}$$

$$F = \frac{(6.67\times10^{-11}\ \text{Nm}^2/\text{Kg}^2)(5.98\times10^{24}\ \text{Kg})(7.35\times10^{22}\ \text{Kg})}{(3.84\times10^{8}\ \text{m})^2}$$

$$F = 1.99\times10^{20}\ \text{N}$$

If we now look at the electrostatic force between the earth and the moon, we encounter a very different situation than that of Example 1. Because nature acts to oppose an imbalance in electrical charge, the net charge on macroscopic objects such as tables, chairs, moons, and planets is effectively zero.

If there is an imbalance in the overall neutral charge of the earth and the moon, it is so small as to be negligible. With no net charge on the two objects in question, Coulomb's Law tell us that there is no electrostatic force between them. Therefore, at this size scale, the gravitational force is by far the dominant force of nature.

Conclusion

In summary, we see that in the subatomic world, the electromagnetic force is much larger and more important than is the gravitational force. In contrast, when we reach the size scale of planets and stars, the electromagnetic force becomes negligible and the gravitational force is the dominant force.

Exercises

1. Calculate the gravitational force between the earth and the sun using the astronomical data found in *College Physics.*

2. Calculate the electrostatic force between two 1 C charges if they are separated by a distance of 3.2×10^{-4} m.

3. Find the net gravitational force on each of the three masses shown in the figure.

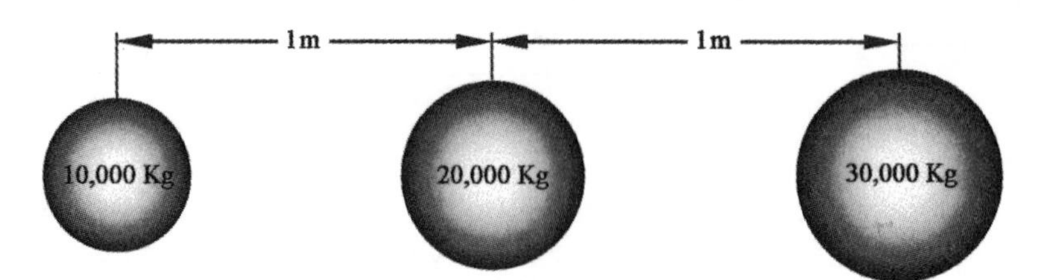

4. Calculate the net force on the 5 C charge.

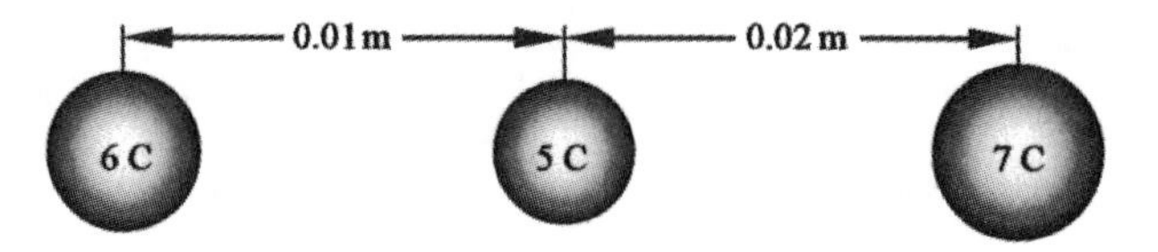

5. Using measurements taken from the surface of the earth, find the point between the earth and the moon at which a mass could be located so that a zero net gravitational force would act on the mass. Use the average earth-moon distance found in *College Physics*.

6. Measured from the 3 C charge, where must an electric charge be placed so that there is no net electrostatic force on it?

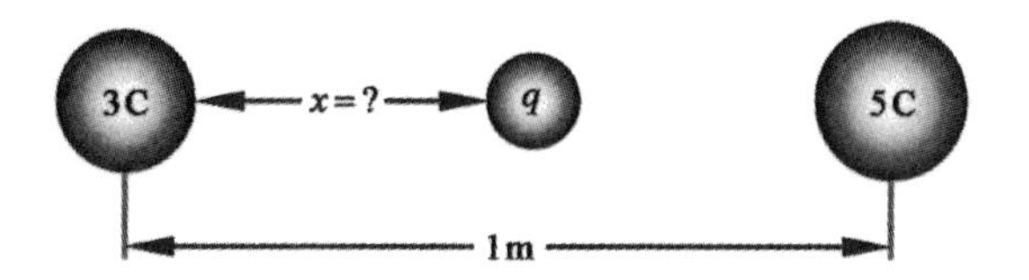

ELECTRONICS 8

Potential Versus Potential Energy

In this application, we explain the electronics concept of voltage using the physical principle of potential energy.

From previous dc electronics studies, we are familiar with basic circuits containing a source voltage, and resistors wired in a variety of combinations like those illustrated in the following figures:

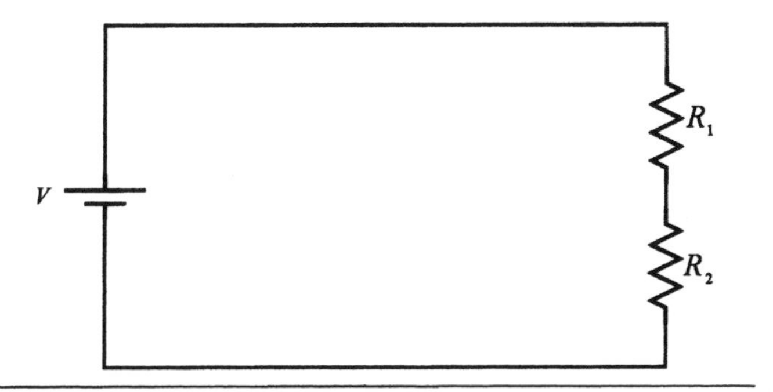

Figure 8.1

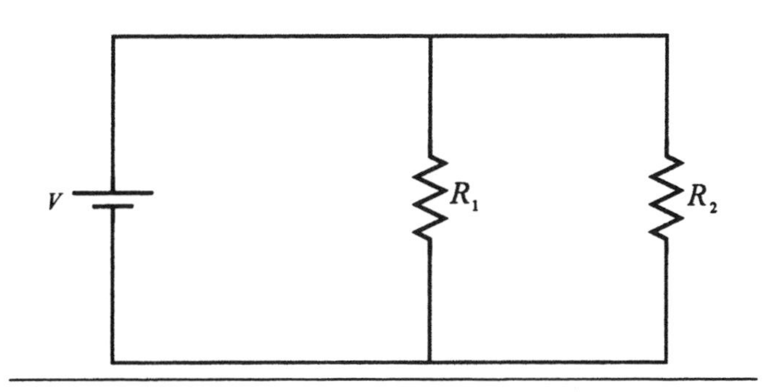

Figure 8.2

Measurements can be taken to find the voltage across a certain resistor, or the total voltage drop along a branch of the circuit, etc. As another example, we can also apply this concept of voltage to more complex circuits like a common-emitter transistor circuit:

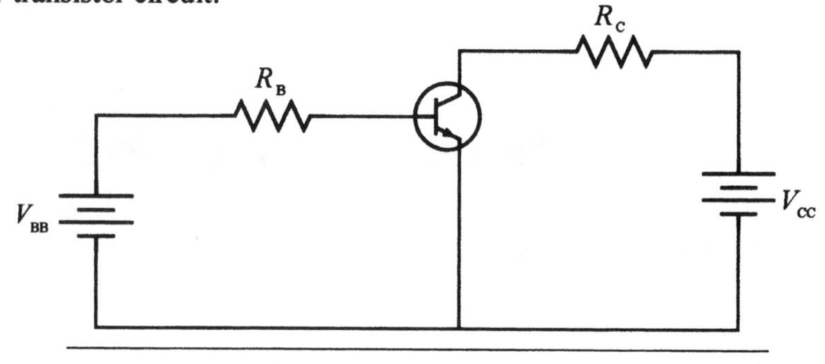

Figure 8.3

The goal of this application is to explain this familiar electronics quantity of voltage in *physical terms*.

Electrostatic Potential

Connection

For a more complete discussion of the force on an electric charge in an electric field turn to Electronics Application 5.

If a test charge, q, is located in an electric field generated by a much larger set of charges, the test charge will experience a force because of the electric field.

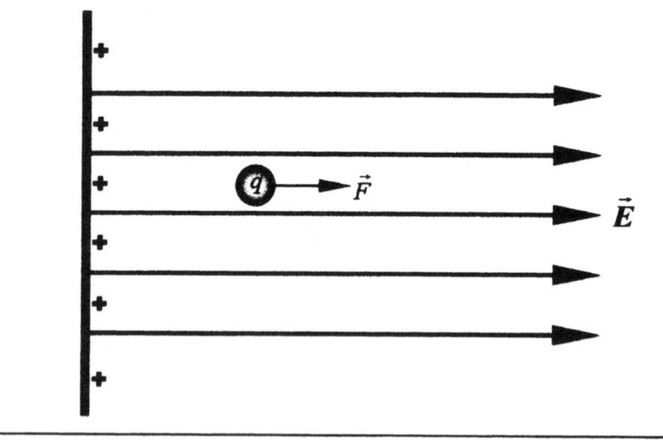

Figure 8.4

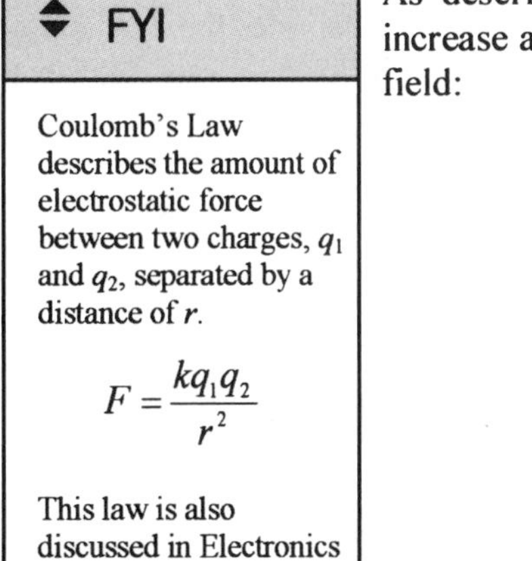

FYI

Coulomb's Law describes the amount of electrostatic force between two charges, q_1 and q_2, separated by a distance of r.

$$F = \frac{kq_1q_2}{r^2}$$

This law is also discussed in Electronics Application 7.

As described by Coulomb's Law, the repulsive force on the test charge will increase as the test charge moves closer to the charge distribution generating the field:

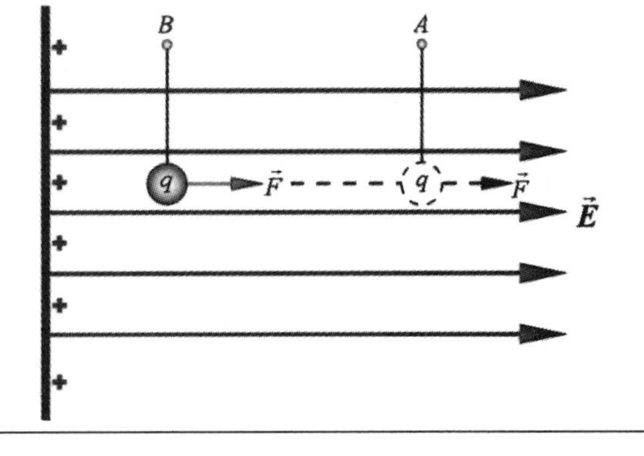

Figure 8.5

Thus, if the charge is released from point B, as in Figure 8.5, it will move to the right with more kinetic energy than if it were released from point A. This additional kinetic energy is a result of the potential energy that is stored by moving the test charge from A to B.

Notice that the potential energy stored when the test charge is located at B is larger than when it is located at A. This situation is analogous to a mass falling from an elevation of B to an elevation of A.

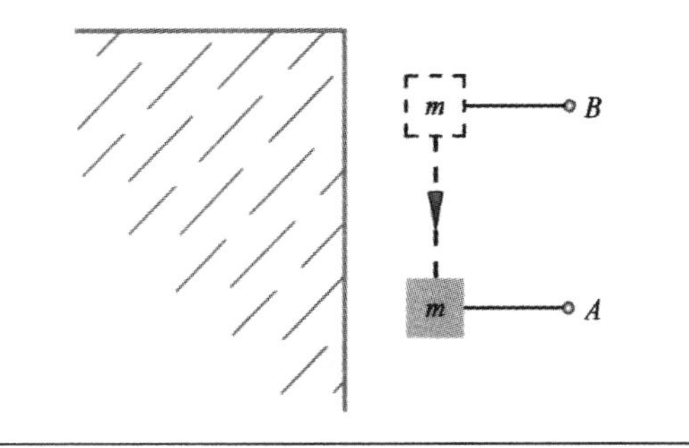

Figure 8.6

Both kinetic and potential energy are presented in detail in Chapter 5 of *College Physics.*

What is important to note is that only movement along an electric field line in Figure 8.5 affects the stored potential energy. If, for example, the test charge moves perpendicular to the field lines, as in Figure 8.7, it does not move closer to or farther from the charge distribution generating the field.

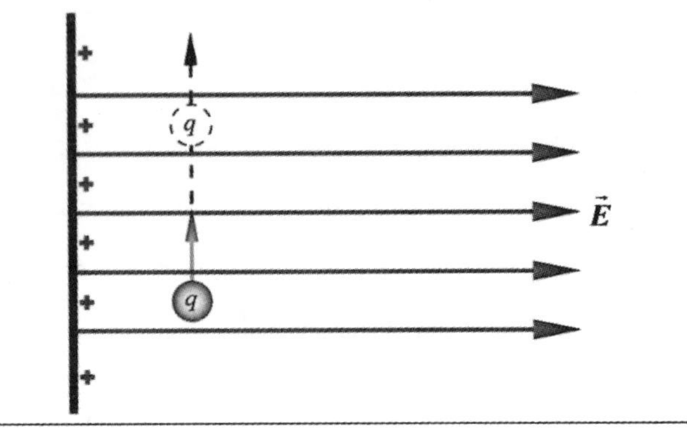

Figure 8.7

The motion of the test charge in Figure 8.7 is analogous to moving a mass horizontally at a height of A. In this case, the amount of stored gravitational potential energy does not change because the mass has not moved along a gravitational field line.

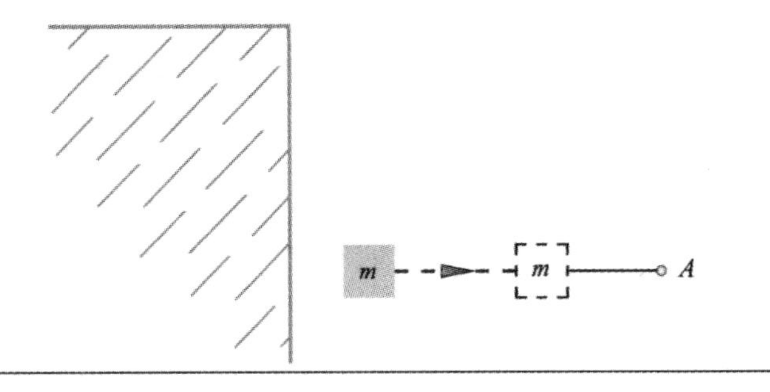

Figure 8.8

Connection

As discussed in Electronics Application 5, the force on a charge in an electric field is given by

$$\vec{F} = q\vec{E}$$

Thus, if an electric charge moves perpendicular to an electric field line, there is no change in the amount of stored potential energy.

Because the size of the test charge affects the amount of force it experiences in the field, and thus the change in the potential energy, it is beneficial to define a quantity that is independent of the size of the test charge being used to evaluate the field. This quantity is known as the *electric potential* (V) and is defined to be the amount of potential energy *per unit charge*:

$$V = \frac{\text{PE}}{q} \quad \textbf{(1)}$$

Since potential is usually measured in *volts*, we commonly refer to it as *voltage*. Because the zero value of potential energy depends upon the point chosen, only the *difference* in potential can be measured physically. The potential, or voltage, V, at a point has no meaning until a zero point, or *ground*, is chosen. Once this zero point is chosen, the potential difference between the point of interest and ground can be calculated.

Consequently, when we say that the voltage across a resistor is V,

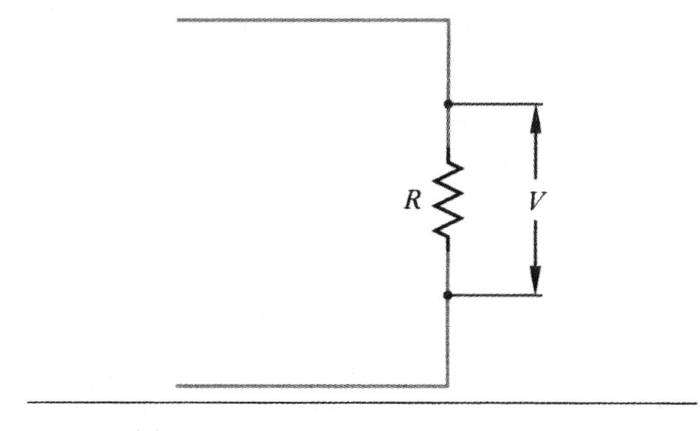

Figure 8.9

we are actually making a statement about the amount of potential energy lost per unit charge as a charge moves through the resistor.

Extending this idea to the common-emitter circuit shown at the beginning of this application, we can now interpret the commonly used notations V_C, V_B, V_E, and V_{CE}.

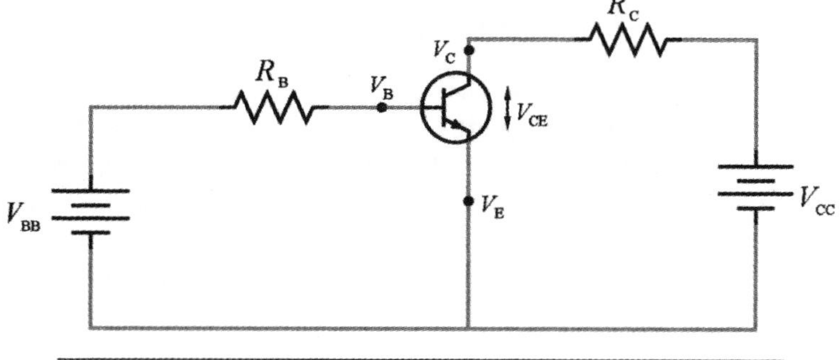

Figure 8.10

Because only differences in potential can be measured, each of these symbols must represent the potential difference between two points in the circuit. Thus, V_{CE} represents the potential difference measured between the collector and the emitter portions of the transistor. The subscripts C and E indicate these two points.

However, the three variables, V_C, V_E, and V_B only have one subscript. In a transistor schematic this notation indicates that the potential difference is being measured between the indicated point and ground.

The Electron-Volt

The definition of potential is useful in explaining the origin of the energy unit of electron-volts (eV) that was introduced in Electronics Application 2. Electron-volts are the principle unit of energy when working with the individual electrons that comprise an electric current.

Beginning with Equation (1) and multiplying both sides by q,

$$\frac{\text{PE}}{q} = V$$

$$\text{PE} = qV \qquad \textbf{(2)}$$

yields an expression for the amount of potential energy possessed by a charge, q, if it is located at a potential of V, measured to ground. Thus, the electron-volt is the amount of potential energy acquired by an electron if it is moved through a potential difference of 1 V. Inserting the magnitude of the charge on an electron, 1.6×10^{-19} C, and the 1 V potential difference into Equation (2) yields the conversion factor between the SI standard unit of joules and electron-volts:

$$1\,\text{eV} = 1.6 \times 10^{-19}\ \text{J}$$

Conclusion

A comprehensive knowledge of the physical principles involved in circuits is central to the study of electronics. This application demonstrates how understanding that the voltage between two points is the difference in the potential energy per unit charge provides a powerful tool for analyzing circuits. This tool is used in Electronics Application 9 to determine the equations for the total resistance in series and parallel circuits.

Exercises

1. Calculate the potential energy of an electron that is located at a potential of 10 V.

2. Measured to ground, an electron has 8.0×10^{-19} J when located at a certain point in a circuit. What is the voltage measured to ground from this point?

3. In the following circuit, find

   a) the voltage at the points A, B, and C
   b) the potential energy of a conduction electron if it is located at each of these points.

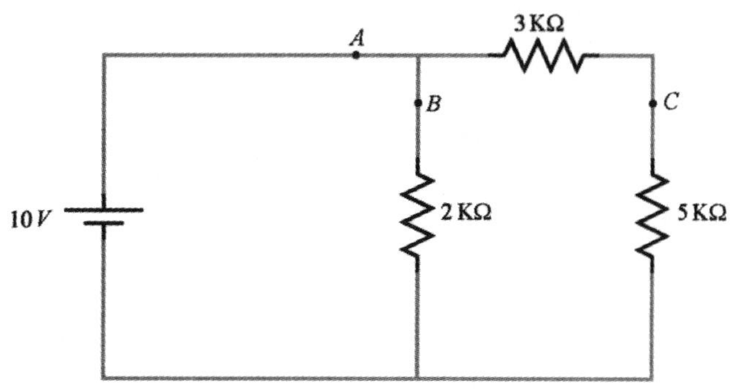

4. In an ac circuit, a conduction electron oscillates according to the frequency of the source voltage. Discuss the potential energy of a conduction electron in such a circuit.

5. In a pn-junction diode, the depletion layer inside the junction has a potential difference across it called the *barrier potential.* Discuss how the material from this application relates to this potential difference inside pn-junction diodes.

6. The *breakdown voltage* for a pn-junction diode is the voltage at which the minority charge carrier current becomes large due to the avalanche effect. Discuss how the avalanche effect and the breakdown voltage are related to the material in this application.

ELECTRONICS 9

Applying the Principles of Energy to Series and Parallel Resistors

In this application, we use the principles of energy to derive the equations for the total resistance of series and parallel circuits.

Circuits containing both series and parallel resistors provide the perfect setting in which to illustrate how an understanding of the underlying *physical* principles deepens and clarifies our understanding of electronics. In this application, we use the idea of an electric charge's potential energy to explain the familiar equations used to find the total resistance of resistors wired in series and in parallel. We will also be able to show the physical origin of Kirchhoff's Voltage and Current Laws familiar from basic electronics.

Potential Energy and Circuits

In Application 8, *Potential Versus Potential Energy*, we learned that an electric charge, q, located at an electrostatic potential, V, possesses potential energy. This potential energy is calculated using the expression

> **Connection**
>
> Potential Energy is discussed in Section 5.4 of *College Physics.*

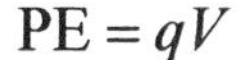

$$\mathrm{PE} = qV$$

The potential used to calculate the potential energy is proportional to the electric field that is present in the wire once the switch in the circuit is closed.

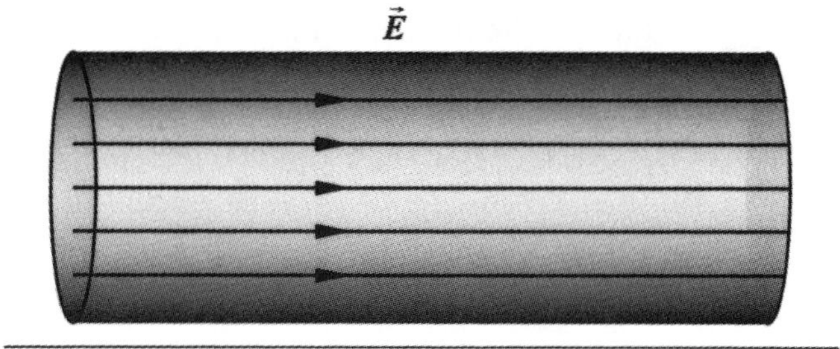

Figure 9.1

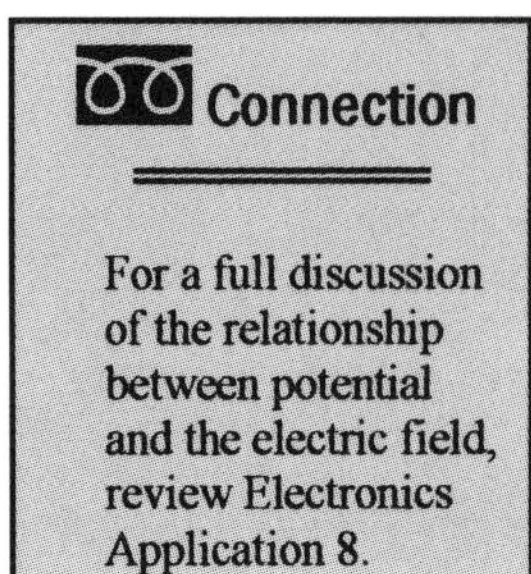

For a full discussion of the relationship between potential and the electric field, review Electronics Application 8.

Although the electric field is assumed constant at all points, the *potential* depends upon the location of a point of interest in the circuit. For example, when we measure the voltage between the two points A and B, we are actually measuring the difference in the value of the electrostatic potential between the points:

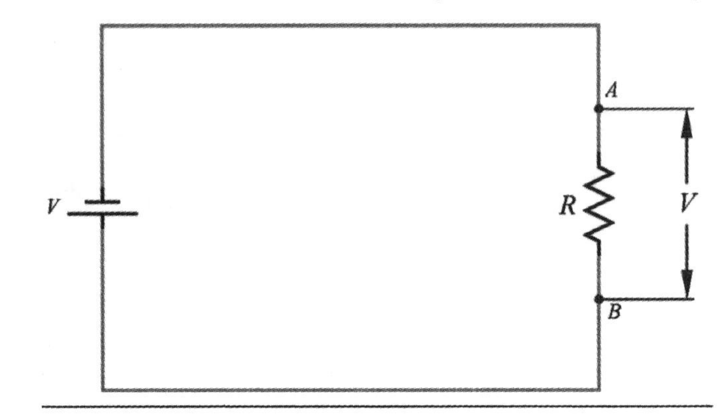

Figure 9.2

Because the potential energy of an electric charge depends upon the potential at which it is located, the potential energy possessed by the charge will be different at points A and B.

Consequently, we see that as the charge moves through the circuit, its potential energy will vary as it is located at points with differing potentials even though its charge remains constant. In other words, if $V_A > V_B$, the charge will lose potential energy by moving from A to B.

Because the voltage (potential) decreases as we move through the circuit, we find the ***maximum*** potential energy that a charge can possess by taking the size of the charge and multiplying it by the source voltage, V:

$$\text{PE}_{\text{max}} = qV$$

Thus, we see that when viewed from the standpoint of potential energy, we can think of the motion of the charges comprising an electric current in the same way as we think of a mass falling from a cliff of height h. Just as the mass moves in the direction that will minimize its potential energy, the individual charges in an electric current "fall" through the circuit in the direction that will minimize their potential energy. Where the amount of starting potential energy for the mass is determined by the height of the cliff, the potential energy of an electric charge is determined by the size of the source voltage.

Kirchhoff's Voltage Law

To construct an explanation for Kirchhoff's Voltage Law based on the principles of energy, we draw on the knowledge that we have gained about how electric charges in a circuit act to minimize their potential energy and the analogy of a falling mass.

> ***Kirchhoff's Voltage Law***
>
> In a closed circuit, the algebraic sum of all the voltages is equal to zero.

In less formal language, Kirchhoff's Voltage Law states that in a closed loop of a circuit, the voltage drops and/or gains across all of the elements must be equal to the source voltage, V.

From the standpoint of potential energy, this statement makes sense. It corresponds to the *physical* fact that by the time a charge makes one full "trip" around the loop, it must have lost all of the potential energy provided to it by the source voltage. Using the falling mass analogy, this "trip" corresponds to the mass hitting the ground. Once the mass is at ground level, it has lost all of its potential energy.

Connection

Conservation of Energy is discussed in Section 5.5 of *College Physics.*

Thus we see that Kirchoff's Voltage Law is the same as the familiar principle of conservation of energy phrased in the language of electronics. Let's now apply these principles to a circuit that contains a source voltage and two resistors wired in series.

Resistors Wired in Series

Suppose that we have the following circuit containing a source voltage and the two resistors, R_1 and R_2, wired in series:

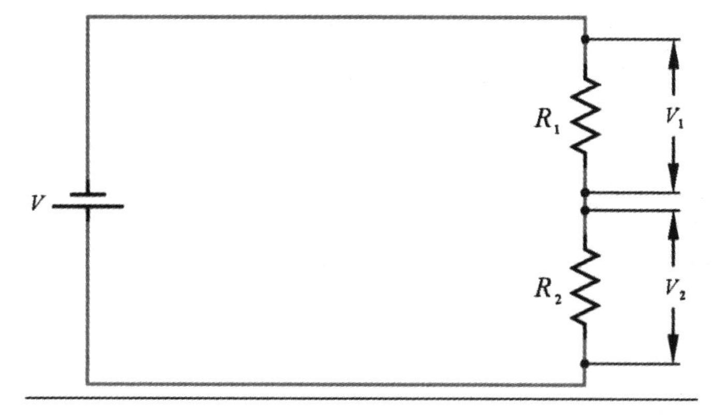

Figure 9.3

The principle of conservation of energy (Kirchhoff's Voltage Law) tells us that when we add the voltage drops across the two resistors, the sum must be equal to the source voltage because these are the only elements in the circuit:

$$V = V_1 + V_2 \qquad \textbf{(1)}$$

Since the voltage drop across a resistor can be expressed as the product of the resistor and the current flow through it,

$$V = IR$$

we can replace both V_1 and V_2 in Equation (1) with terms that contain the value of the resistor and the amount of current flowing through the resistor. Thus Equation (1) can be rewritten as

$$V = IR_1 + IR_2 \qquad \textbf{(2)}$$

Notice that although we placed subscripts on the resistors in Equation (2) to distinguish between them, we did not do so for the current flow, I. This step is unnecessary because, by definition, two resistors wired in series have identical amounts of current flowing through them.

Beginning with Equation (2) and factoring out the current, I,

$$V = IR_1 + IR_2$$
$$V = I(R_1 + R_2)$$

yields the result for series resistors. Since V is the total voltage in the circuit and I is the total current flow, the expression $R_1 + R_2$ must be the total resistance in the circuit. Thus, to find the total resistance of two resistors wired in series, we simply add them together:

$$R_{\text{Total}} = R_1 + R_2$$

Resistors Wired in Parallel

The principle of conservation of energy and our analogy of a mass in a gravitational field can also be applied to a circuit containing two resistors wired in parallel like the one in the figure:

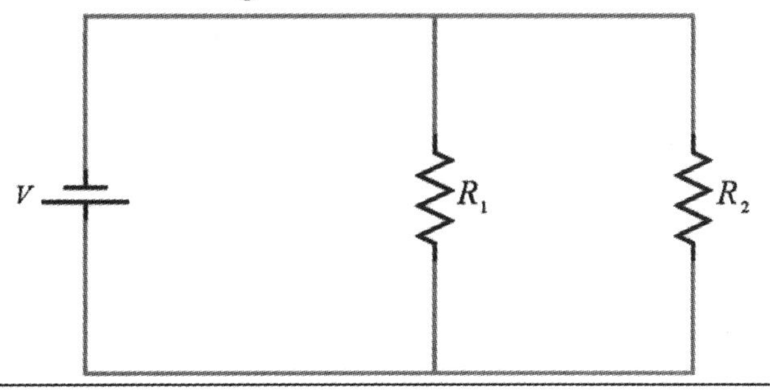

Figure 9.4

Our analysis of this circuit can be simplified by drawing on Kirchhoff's Voltage Law and thinking of the circuit as two individual loops:

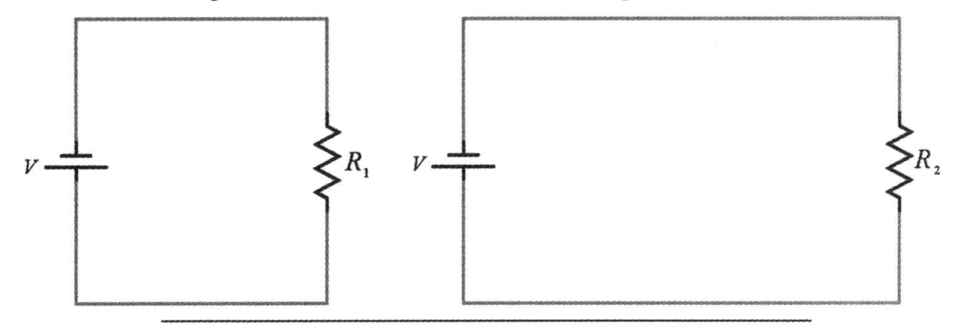

Figure 9.5

As in the case of the series circuit, it is actually the principle of potential energy that allows us the freedom to address the loops independently. Returning to the idea of a mass in a gravitational field helps us to visualize this concept.

Suppose that a worker's job requires him to lift balls of identical mass to a height, *h*, and roll them toward two tubes of water, as in the following diagram:

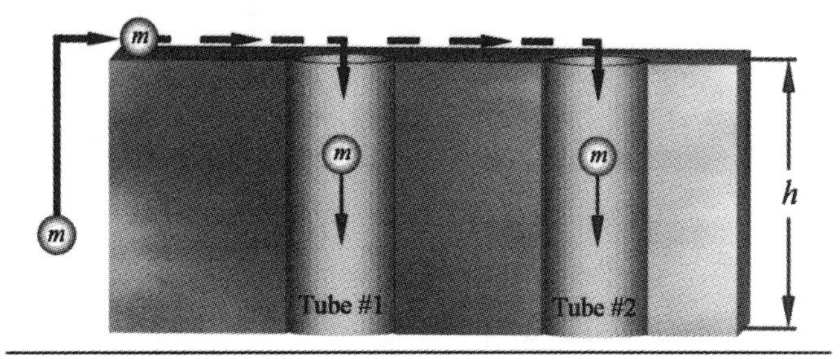

Figure 9.6

Because the worker is good at his job, he rolls so many balls that not all of them are able to fall down tube #1 and some are forced to fall down tube #2. However, regardless of the tube taken, when a ball hits the ground it has lost all of the potential energy given to it by the worker.

We can apply the same principles to our parallel circuit. In this case, the source voltage acts as the worker and the resistors function as the tubes.

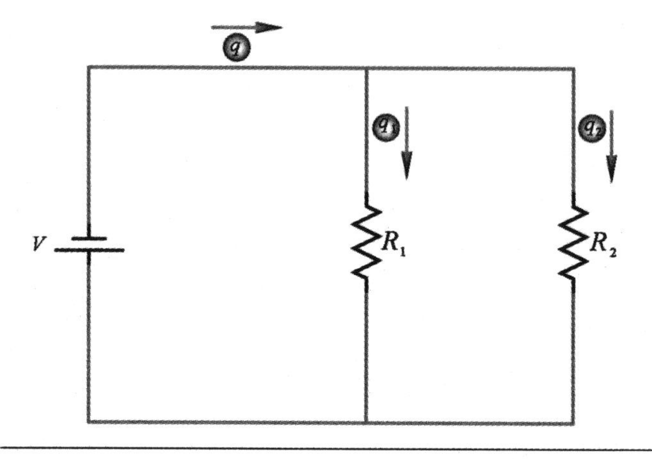

Figure 9.7

Each electric charge in the current is given a certain amount of potential energy by the source voltage. Regardless of whether the charge moves "down" resistor #1 or #2, by the time it hits "bottom" it must have exhausted all of its potential energy.

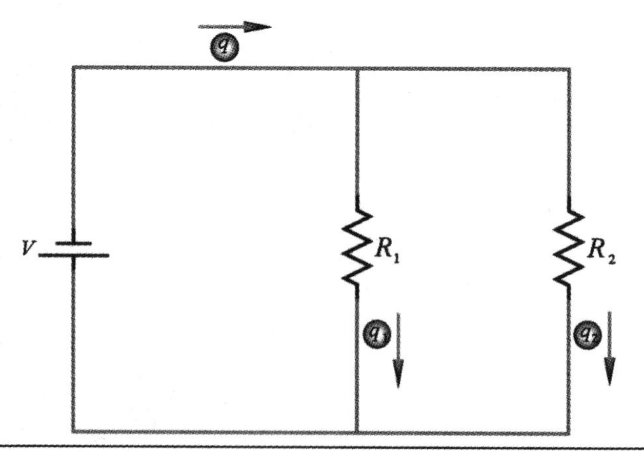

Figure 9.8

> **Connection**
>
> The concept of friction is introduced in Section 4.6 of *College Physics.*
>
> The role friction plays in decreasing the energy of a system is discussed in Section 5.5

The analogy between the circuit and the mass system can be extended by realizing that both systems are dissipating energy in the form of heat. Just as friction between the surface of the balls and the water molecules in the tubes produces heat that will raise the temperature of the water, each resistor heats up as the current flows through it. This dissipation of thermal energy is a result of friction between the charges and the atoms and molecules comprising each resistor and provides a springboard for examining Kirchhoff's Current Law.

Kirchhoff's Current Law

If we return to our example of balls and tubes, we see that although each ball loses all of its potential energy, we do not lose any balls. If we add together the number of balls moving down tube #1 and tube #2, the sum must equal the total number of balls rolled by the worker.

Similarly, if we add together the number of charges moving through resistor #1 and resistor #2, we will have the total number of charges that entered the node *A*:

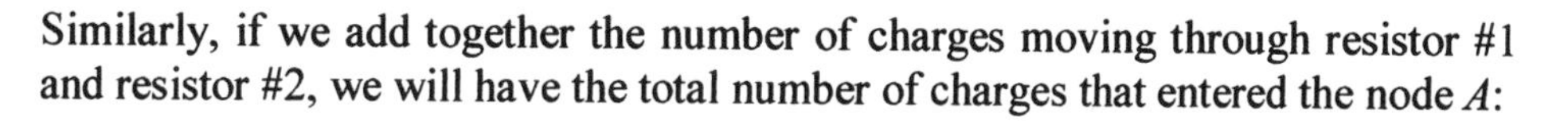

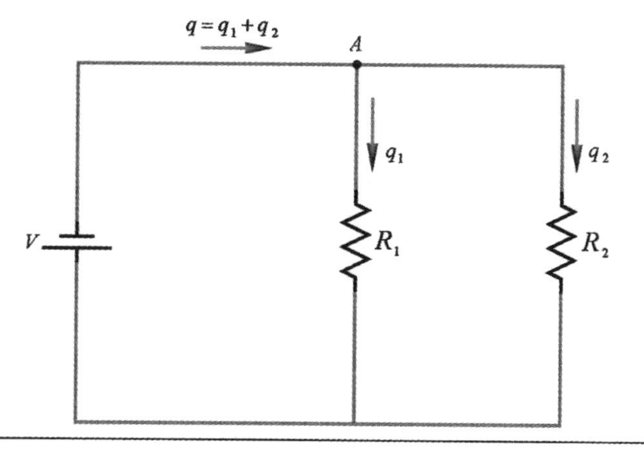

Figure 9.9

Using the term *current* instead of charges, we arrive at Kirchhoff's Current Law:

> ***Kirchhoff's Current Law***
>
> The sum of all currents leaving a circuit node must be equal to the sum of all the currents entering the node.

In equation form this law is expressed as

$$I = I_1 + I_2$$

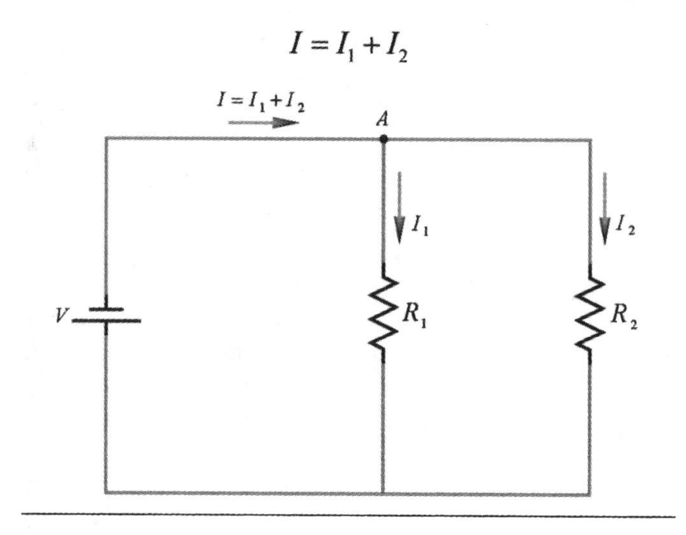

Figure 9.10

▸ The Total Resistance of Two Resistors in Parallel

In the previous two sections, we explained Kirchhoff's Current Law and Voltage Law using the language of potential energy. Now we will use these laws along with the concept of potential energy to derive the equation used to find the total resistance of two resistors wired in parallel.

Consider the circuit shown in the figure:

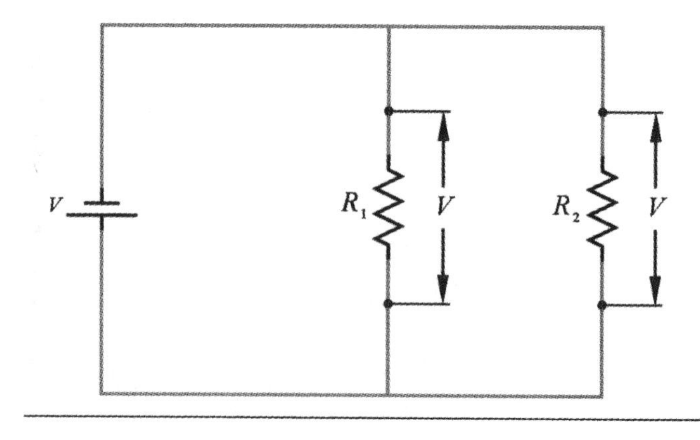

Figure 9.11

From the section on Kirchoff's Voltage Law, we know that if we trace around a closed loop in the circuit, the algebraic sum of all the circuit elements must be zero. Also, we know from the section on parallel resistors that the voltage across two resistors in parallel must be the same. We can therefore apply Ohm's Law to each resistor in the figure, yielding

$$V = I_1 R_1$$

and

$$V = I_2 R_2$$

We can rewrite these two equations in more useful forms as

$$I_1 = \frac{V}{R_1}$$

and

$$I_2 = \frac{V}{R_2}$$

When we insert the two new expressions into the current conservation expression, or Kirchhoff's Current Law,

$$I = I_1 + I_2$$

we obtain the equation

$$I = \frac{V}{R_1} + \frac{V}{R_2}$$

Next, we factor out the voltage, V,

$$I = V\left(\frac{1}{R_1} + \frac{1}{R_2}\right)$$

and then divide both sides by the term involving the resistors:

$$\frac{I}{\frac{1}{R_1} + \frac{1}{R_2}} = V$$

This expression can be rewritten as

$$I\left(\frac{1}{\frac{1}{R_1}+\frac{1}{R_2}}\right)=V$$

Since V is the total voltage, and I is the total current flow, we know that the term

$$\frac{1}{\frac{1}{R_1}+\frac{1}{R_2}}$$

must be equal to the total resistance in the circuit; thus, the equation for the total resistance of two resistors in parallel is

$$R_{\text{Total}}=\frac{1}{\frac{1}{R_1}+\frac{1}{R_2}} \qquad \textbf{(3)}$$

Equation (3) is usually written in the more concise form

$$\frac{1}{R_{\text{Total}}}=\frac{1}{R_1}+\frac{1}{R_2}$$

The expression for the total resistance can be extended to more than two resistors wired in parallel.

Three resistors:

$$\frac{1}{R_{\text{Total}}}=\frac{1}{R_1}+\frac{1}{R_2}+\frac{1}{R_3}$$

Four resistors:

$$\frac{1}{R_{\text{Total}}}=\frac{1}{R_1}+\frac{1}{R_2}+\frac{1}{R_3}+\frac{1}{R_4}$$

Conclusion

The principles of potential energy and conservation of energy are powerful tools for circuit analysis. Once we realize that the much deeper principle of energy conservation is at the center of Kirchhoff's Laws, their origin and meaning become clearer. Although every electronics student learns these laws early in his or her career, most do not realize that these laws are based on the same physical principles that describe falling masses and many other non-electronic situations.

Exercises

1. Find the total resistance in the following series circuit:

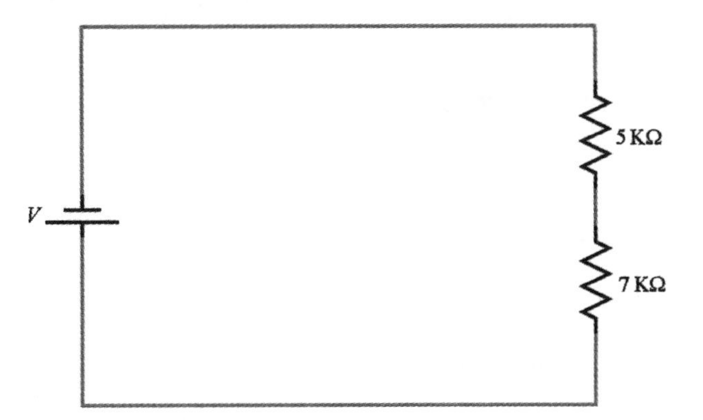

2. Find the total resistance in the following parallel circuit:

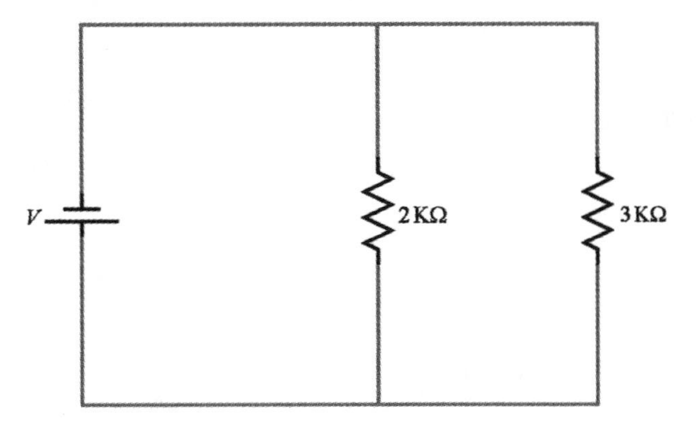

3. Find the total resistance in the following series-parallel circuit:

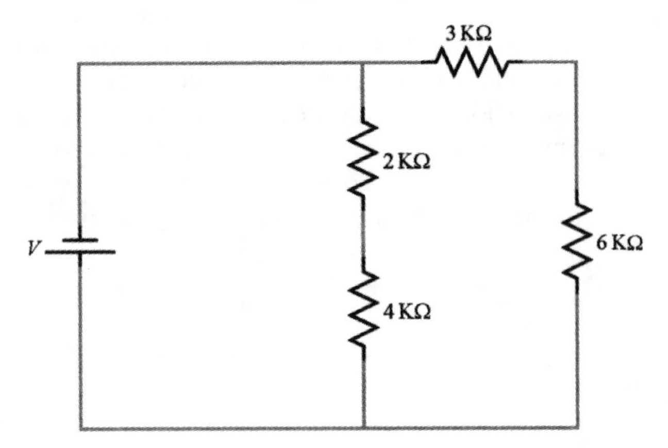

4. What is the voltage across each of the resistors in Exercise 3 if there is a source voltage of 10 V?

5. Find the current flow at point *A*.

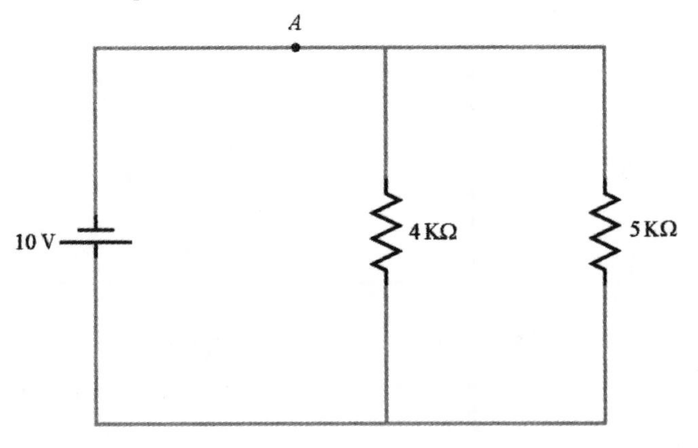

6. Find the current flow at point *A*.

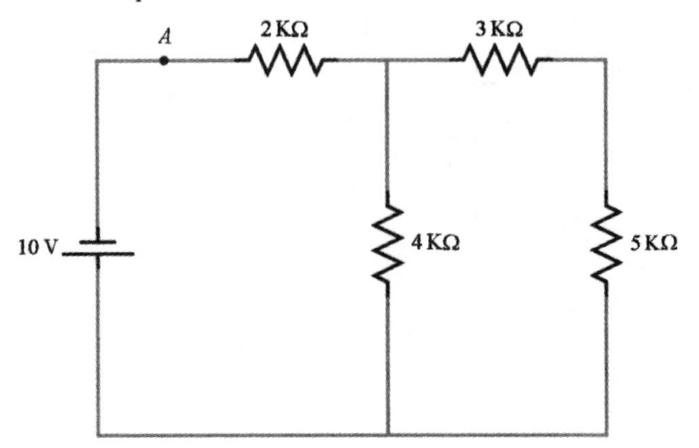

7. From basic electronics, the power dissipated by a resistor is given by

$$P = I^2 R$$

Use this equation, along with the material in this application, to find the current flow through a 10 KΩ resistor that is dissipating 0.25 W of power.

8. ***Challenge Problem***

Discuss how the concepts of potential and potential energy explain a familiar electronics diagnostic tool, the Wheatstone Bridge.

ELECTRONICS 10

Heat, Kinetic Energy, and the Breakdown of PN-Junction Diodes

In this application, we use the principles of heat and kinetic energy discussed in *College Physics* to give a physical explanation of the breakdown of pn-junction diodes.

Electronics Review

In a p-type semiconductor, the majority charge carriers are holes, while in an n-type semiconductor, the majority charge carriers are electrons.

Connection

The depletion layer is also discussed in Electronics Application 5 - *The Force on an Electron in an Electric Field*.

The standard pn-junction diode is a device made from two types of semiconductors, p-type and n-type. When p-type and n-type semiconductors are used together, they form an electronic device known as a diode that allows electric current to flow in one direction only. If we try to force a current flow in the wrong direction, the diode suffers *breakdown*. In this application, we will give a physical explanation of the breakdown of a pn-junction diode, which is familiar from solid-state electronics.

The Depletion Layer and Biasing

The interface between the two types of semiconducting materials inside the diode is called the pn-junction. Inside this junction is a region that has been depleted of mobile charge carriers called the *depletion layer*. This layer is formed when electrons from the n-type semiconductor fill holes in the p-type semiconductor. Because the n-side of the diode loses electrons, it becomes more positive. Conversely, the p-side becomes more negative because of the acquisition of additional negative charges.

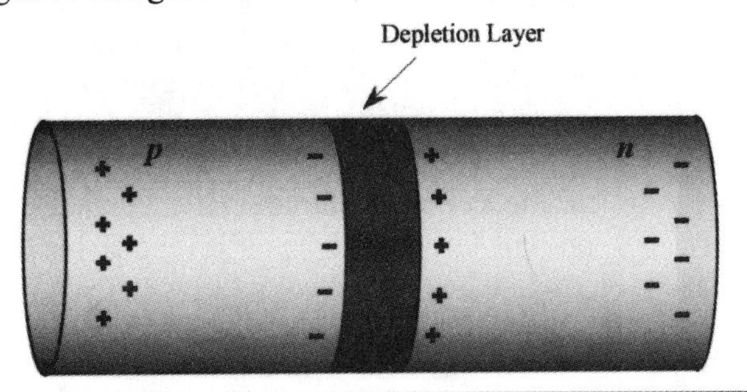

Figure 10.1

If the diode is wired to an external voltage, the depletion layer will widen or shrink depending upon the polarity of the applied voltage.

In Figure 10.2a, the applied voltage will cause the depletion layer to shrink until it disappears allowing current to flow through the diode. In this case, the diode is said to be *forward biased.*

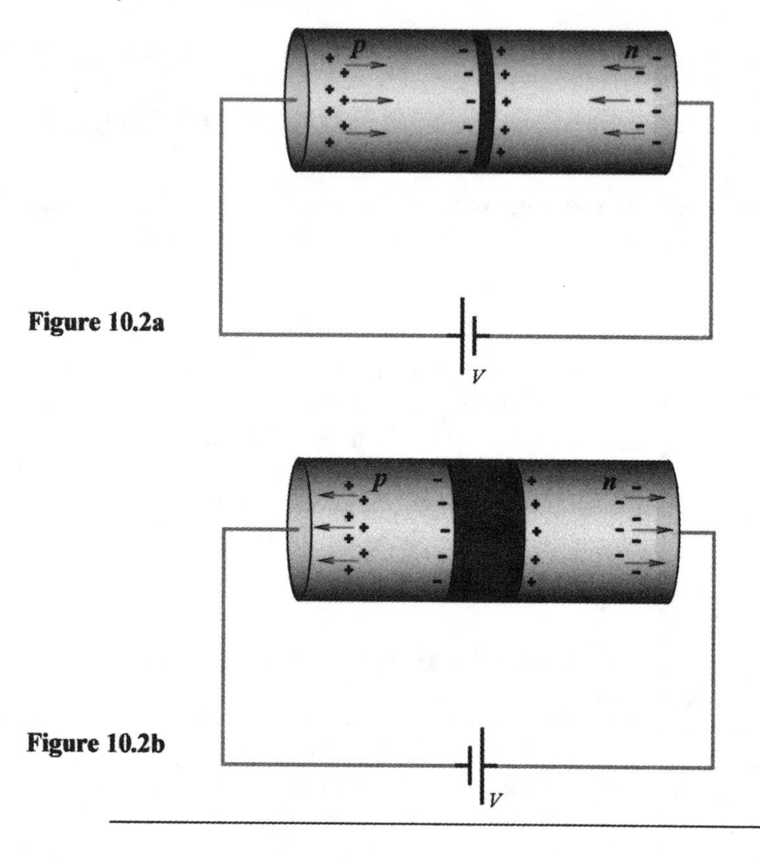

Figure 10.2a

Figure 10.2b

If the polarity of the applied voltage is reversed, as in Figure 10.2b, the negative terminal of the voltage source attracts positive charges away from the junction and the positive terminal attracts negative charges away from the junction. This causes the depletion layer to grow and prevents current from flowing through the diode. In this case, the diode is said to be *reverse biased.* Armed with this background information about pn-junction diodes, we can now draw on a number of the concepts introduced in *College Physics* to explain the behavior of a pn-junction diode when it is reverse biased.

Minority Carrier Current

Although we have identified the diode in our example as being reverse biased, a small current does, in fact, still flow through it because of thermal effects in the diode. The heat generated inside the diode, caused by its being wired to a source voltage, increases the kinetic energy of the atoms in the diode, producing free electrons and holes throughout the diode. The electrons produced in the p-side of the diode and the holes produced in the n-side are referred to as *minority charge carriers* since they are fewer in number and possess opposite charge to that of the majority charge carriers for each side. Because the electrons and holes produced by heat are minority charge carriers, the majority charge carriers on each side quickly neutralize these free charges. However, when heat causes an electron-hole pair to be created inside the depletion layer, an electron may migrate into the p-side of the diode and move into the external portion of the circuit. In a similar manner, the hole that was also generated can move through the n-side and out the other end of the diode. In each case, the charges that move through each side of the diode are not the majority charge carriers for that side. Thus, there is a small current that flows even when a diode is reverse biased. This current is called *minority carrier current.*

Kinetic Energy and the Avalanche Effect

Connection

Kinetic energy is discussed in Section 5.3 of *College Physics.*

Understanding that heat is the cause of minority carrier current, we can now see how pn-junction diodes suffer breakdown. If the size of the reverse-biasing voltage is increased, the kinetic energy of the electrons that are produced in the junction increases. The larger the source voltage, the faster these thermally produced electrons move into the p-side of the diode.

As these high-energy electrons collide with atoms in the p-side of the diode, they free more electrons. When a high-energy electron collides with an atom, it frees one valence electron from the atom and moves to collide with another atom. Because this newly freed electron also has a large kinetic energy, it will collide with another atom, causing another electron to be freed. The result is a geometric progression as one free electron becomes two, two become four, four become eight, and so forth.

This runaway progression is known as the *avalanche effect* and the reverse voltage required to cause it is called the *breakdown voltage.* The large reverse current produced by the avalanche effect causes the diode to conduct heavily in the reverse direction and the diode suffers *breakdown.*

Conclusion

Understanding the physical properties at play inside electronic devices deepens our understanding of the devices and their operation. In this application, we learned that an enormous number of collisions and other events happen at the atomic level when we reverse bias a diode. Although the breakdown of a diode appears nearly instantaneous in a laboratory setting, in actuality, the principles of heat and kinetic energy affect millions of atoms inside the diode. Once again we see how the laws of *physics* govern the behavior of electronic circuit elements.

Exercises

1. When forward biasing a diode, will any amount of source voltage cause current flow in the diode? Why or why not?

2. How does the discussion of breakdown given in this application relate to zener diodes?

3. An npn-transistor is a "sandwich" of n- and p-type semiconductors. This sandwich results in two depletion layers inside the diode. During normal operation, one of the depletion layers is forward biased while the other is reverse biased. Discuss how the concepts of biasing and breakdown presented in this application apply to the functioning of an npn-transistor.

ELECTRONICS 11

The Power Dissipated in an RLC-Circuit

In this application we use the physical principles of friction, heat, and energy to find the power dissipated by an RLC-circuit.

> **Connection**
>
> The concept of power is discussed in Section 5.6 of *College Physics*.

Power can be defined as the rate at which the energy in a system changes:

$$P = \frac{\Delta E}{\Delta t}$$

As applied to our example RLC-circuit, this means that a circuit element can only dissipate power if it dissipates energy. Because heat is a form of energy, if a circuit element does not generate any heat, it cannot dissipate any energy. Hence, without a changing energy the circuit element cannot dissipate any power. In this application we will analyze each element in the RLC-circuit to determine where the energy in the circuit is actually being lost. Our starting point for this analysis is the overall impedance in the circuit. By analyzing *physically* the impedance of a series RLC-circuit, like the one illustrated in Figure 11.1, we can gain a better understanding of how and where energy is being lost in the circuit and the physical principles that come into play.

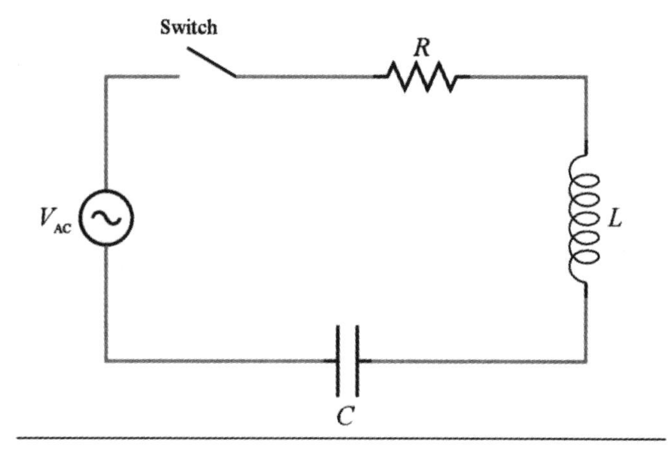

Figure 11.1

Impedance

From basic ac electronics we know that the alternating voltage, V_{AC}, the resulting current flow, I, and the impedance to the current flow, Z, are all related by an equation from ac theory which is similar to Ohm's Law:

$$V_{AC} = IZ \qquad (1)$$

For us to understand how and why power is dissipated, we must analyze the impedance in the circuit in greater detail.

The impedance in the circuit results from individual contributions from the resistor, the inductor, and the capacitor. Each one of these circuit elements provides opposition to the current flow. However, the *physical* reasons why each circuit element opposes the current flow differ vastly from element to element. Let's look at each of the circuit elements individually.

The Resistor

Although we usually draw a simple diagram such as Figure 11.2 when we draw schematics,

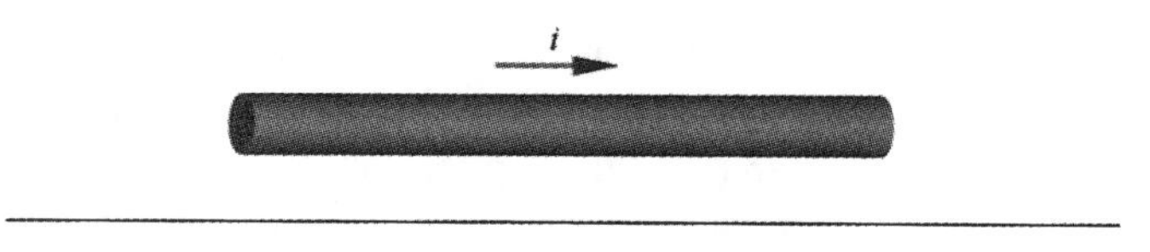

Figure 11.2

Connection

The path of conduction electrons is discussed in Application 4 - *Vectors and Electronics.*

the actual path of the current inside a wire or a resistor is far more complex. A more accurate diagram of the path of a conduction electron is pictured in Figure 11.3:

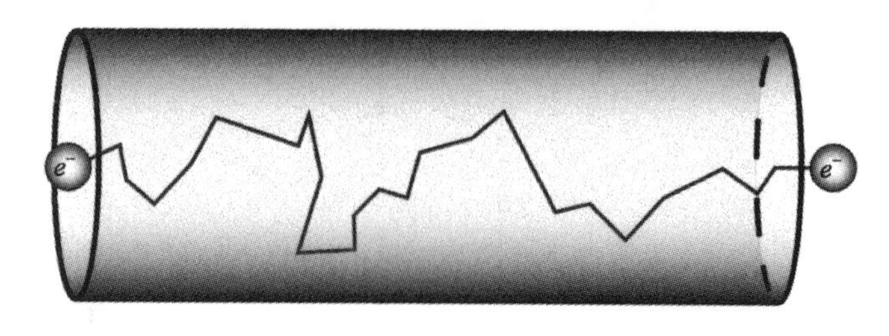

Figure 11.3

As the individual electrons that comprise the current attempt to move through the resistor, they meet with opposition that depends upon the type of material, the temperature of the material, the length of the resistor, and its cross-sectional area. It is reasonable that both the material itself and its temperature affect the resistance in the resistor. Different materials have different densities and molecular structures that affect the ease with which a conduction electron can pass through the resistor.

Connection

Kinetic energy is discussed in Section 5.3 of *College Physics.*

Temperature and heat are discussed in Chapters 10 and 11.

Further, the temperature of the material affects the kinetic energy of the molecules. The higher the temperature of the material, the greater the kinetic energy and vibrational frequency of the material's atoms and molecules. This movement of the material's atoms impedes the flow of the conduction electrons that form the current.

Both the type of material out of which the resistor is constructed and its temperature are incorporated into the *resistivity* of the material (ρ). Resistivity is calculated using

$$\rho = \rho_0 + \rho_0 \alpha (T - T_0)$$

where

- ρ is the resistivity of the material at the temperature, T
- ρ_0 is the resistivity of the material at room temperature, T_0
- α is the temperature coefficient of resistivity of the material
- T is the temperature of the material
- T_0 is room temperature

In the resistivity equation, the units on the temperature coefficient of resistivity must cancel the units on the temperature term. Thus, if the units on the temperature are °C, the units on the coefficient of resistivity will be 1/°C. For a discussion of various temperature scales, see Section 10.2 of *College Physics.*

Notice that the second term in the resistivity equation becomes zero if the material is operating at room temperature. At room temperature, the resistivity becomes simply the baseline resistivity, ρ_0. As the temperature of the material increases above room temperature, the resistivity of the material grows. In electronics, this is normally referred to as a positive temperature coefficient.

We can couple the resistivity of the material with the other two physical constraints on the resistor, its length (L) and its cross-sectional area (A), using the equation

$$R = \rho \frac{L}{A}$$

The length of the resistor increases the resistance because it increases the number of atomic and molecular collisions that the conduction electrons must encounter as they attempt to move through the resistor. Conversely, the cross-sectional area of the resistor actually decreases the resistance of the resistor. As the cross-sectional area of the resistor becomes larger, the distances between the atoms and molecules that comprise the resistor increase. This increased atomic spacing allows for more navigational space as the conduction electrons attempt to move through the resistor.

Connection

Friction is discussed in Section 4.6 of College Physics.

Two results of the atomic collisions experienced by the conduction electrons are friction and heat. These two quantities will be important to us later when we discuss the total power dissipated by the circuit.

In summary, we see that the resistor's contribution to the total impedance of the circuit arises from the atomic and molecular collisions that the current experiences as it passes through the resistor.

The Capacitor

A typical capacitor is constructed by placing two conducting surfaces in close proximity to one another:

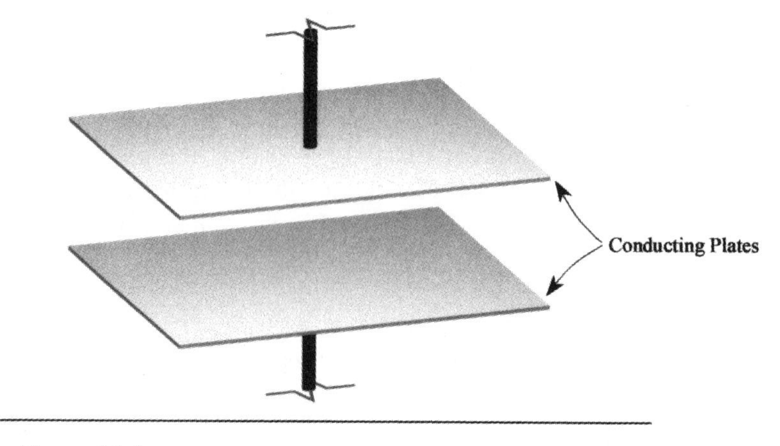

Figure 11.4

Although there are many possible geometric configurations for these two surfaces ranging from cylindrical capacitors to spherical capacitors, etc., it is sufficient for us to focus on the simplest type, a parallel-plate capacitor.

When the two plates are connected to a source voltage, charges begin to accumulate on the plates of the capacitor:

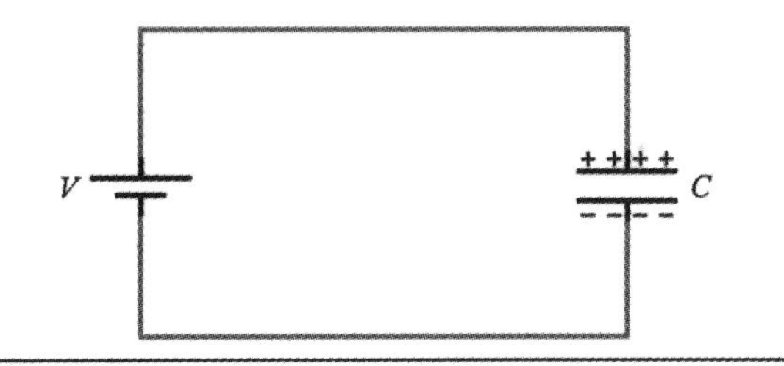

Figure 11.5

These charges cause a potential difference (voltage) to appear across the plates, and more importantly for our purposes, an *electric field* to appear between the plates of the capacitor:

FYI

It should be noted that there are "fringing" effects of the electric field around the edges of the capacitor. For our purposes, it is sufficient to ignore these effects and assume that all electric field lines point straight from the positive plate to negative plate of the capacitor.

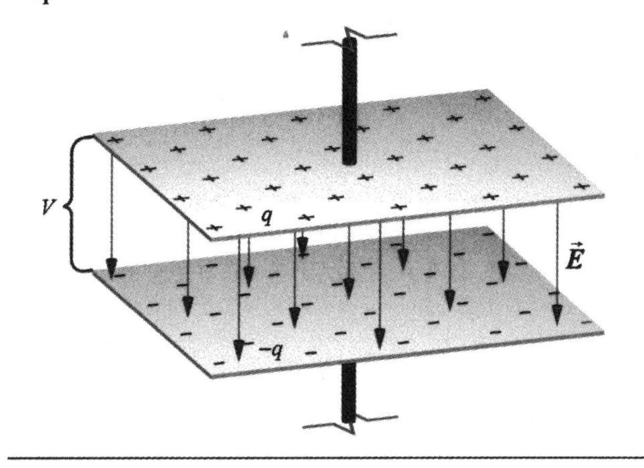

Figure 11.6

This electric field has an orientation that depends upon the polarity of the charges that are stored on the plates of the capacitor. Because electric fields point from positive to negative, we can have either of two possible configurations:

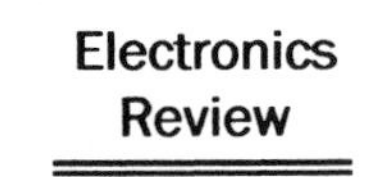

The capacitive reactance of a capacitor is found using

$$X_C = \frac{1}{2\pi fC}$$

where f is the frequency of the source voltage.

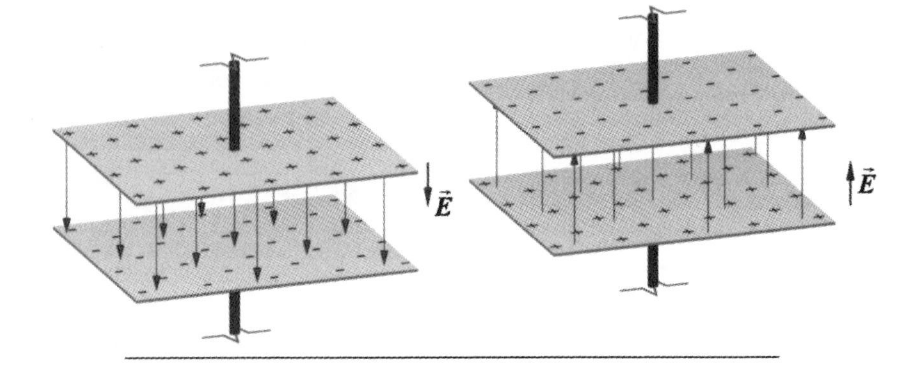

Figure 11.7

Because the capacitor in our RLC-circuit is connected to an alternating voltage source, the polarity of the charges stored on the plates of the capacitor change at the same frequency as the driving frequency of the source. In turn, the electric field generated by the stored charges changes its polarity with a regular frequency.

FYI

The imbalance in the stored charge on the plates of the capacitor also plays a role in generating the capacitive reactance.

Unfortunately, nature opposes changes in electric fields. This ever-changing electric field inside the capacitor produces a quantity known as *capacitive reactance* (X_C). This capacitive reactance acts to oppose the current flow. However, because no friction is produced as the massless electric field changes polarity, little or no heat is generated inside the capacitor.

The Inductor

An inductor is an electronic device that uses the magnetic field produced from a current-carrying wire. As the current flows through the wire, a magnetic field is produced around the wire, the direction of which corresponds to the direction of the current flow:

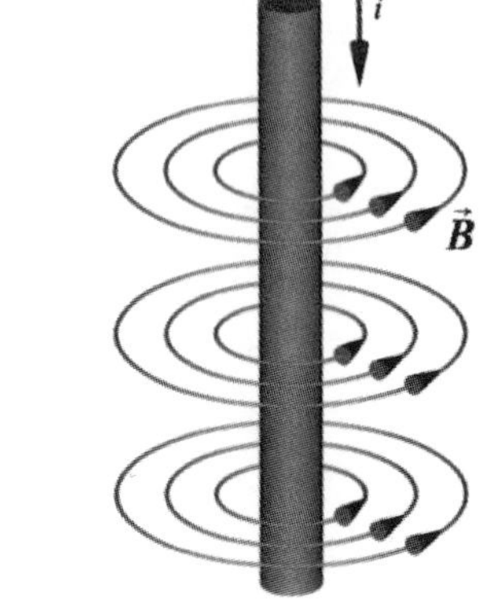

Figure 11.8

If the wire is now wound into the geometric form of a coil, or *solenoid*, the magnetic field generated by the current flow takes on a new form. Because the magnetic field lines that surround the wire have a polarity, a cancellation occurs along the edge of the inductor between the windings. Where one wire has magnetic field lines that point outward, the wire immediately adjacent to it has magnetic field lines that point inward:

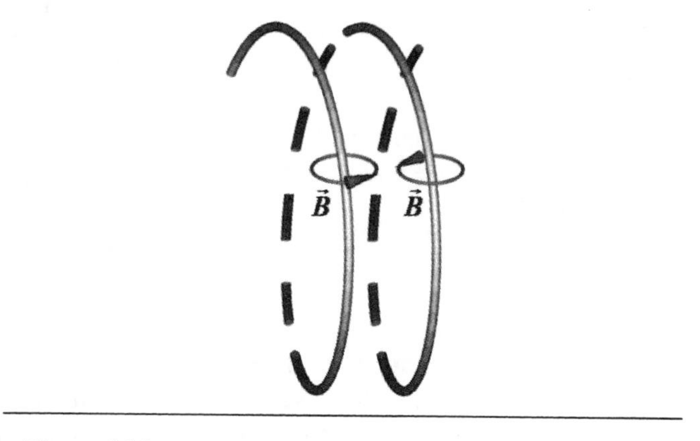

Figure 11.9

The net result is that the only appreciable surviving magnetic field points down the center of the inductor:

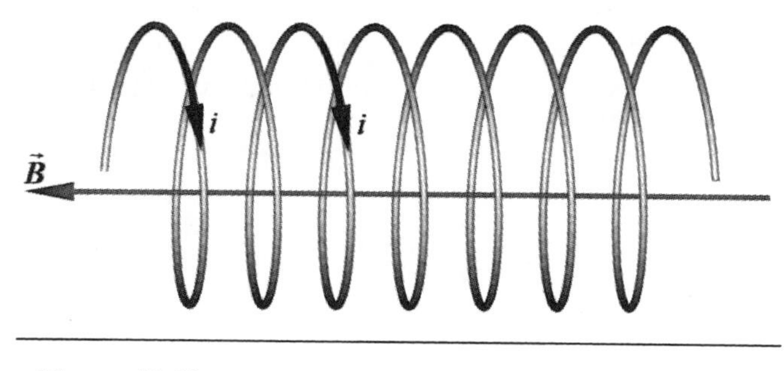

Figure 11.10

Because the inductor in the RLC-circuit is wired to an alternating voltage, the current flow through the inductor changes direction with a frequency determined by the voltage source. Thus, the magnetic field generated by current flow will also change with a regular frequency. Just as in the case of the capacitor, nature resists this change in magnetic field polarity. A quantity known as *inductive reactance* (X_L) opposes the current flow. This inductive reactance provides the inductor's contribution to the overall impedance of the circuit. However, as with the absence of friction or heat loss in the changing electric field of the capacitor, the changing massless magnetic field inside the inductor generates little or no heat.

Electronics Review

The inductive reactance of an inductor is found using

$$X_L = 2\pi f L$$

where f is the frequency of the source voltage.

The Power Dissipated by an RLC-Circuit

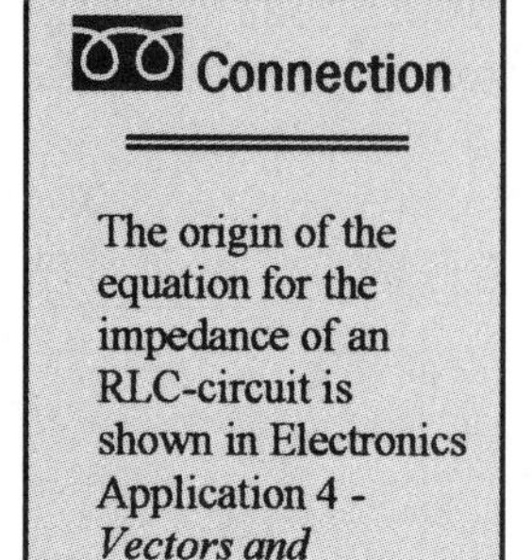

Thus we see that the total impedance of the circuit arises from three sources, the resistance of the resistor, the capacitive reactance of the capacitor, and the inductive reactance of the inductor. This total impedance can be expressed in equation form as

$$Z = \sqrt{R^2 + (X_{\mathrm{L}} - X_{\mathrm{C}})^2}$$

The total current flow in the RLC-circuit can be found by first calculating the impedance and then rearranging Equation (1) to solve for the current flow:

$$\frac{V_{\mathrm{AC}}}{Z} = I$$

However, an interesting situation arises when we attempt to calculate the total power dissipated by the circuit. Although we use the total impedance, Z, to calculate the current flow, we need only use the resistance of the resistor to calculate the power dissipated. Because the capacitor and the inductor do not dissipate any heat, they are not dissipating any energy. Consequently, we are usually able to ignore the reactance provided by both these elements when we calculate the power dissipated by the circuit.

Because we are not taking into account the inductive reactance or the capacitive reactance, the general impedance equation

$$Z = \sqrt{R^2 + (X_{\mathrm{L}} - X_{\mathrm{C}})^2}$$

simplifies to

$$Z = \sqrt{R^2 + (X_{\mathrm{L}} - X_{\mathrm{C}})^2}$$
$$Z = \sqrt{R^2}$$
$$Z = R$$

Thus, instead of using an equation of the form

$$P = I^2 Z$$

to calculate the dissipated power, we can simplify it to

$$P = I^2 R$$

Because the capacitor and the inductor in the circuit do not produce heat as their internal fields change polarities, they do not dissipate any power. Therefore, the only circuit element of the three that must be included in the power calculation is the resistor because of the heat produced by the friction between the conduction electrons of the current and the atoms of the wire.

▸ Conclusion

Once again we see that a firm understanding of the *physical* quantities involved in a circuit deepens our understanding of the electronics principles. In this application, we were able to use a much simpler form of the power equation by understanding the physical quantities of heat, energy, and power that were governing the behavior of the circuit elements.

Exercises

1. Find the power dissipated by the resistor.

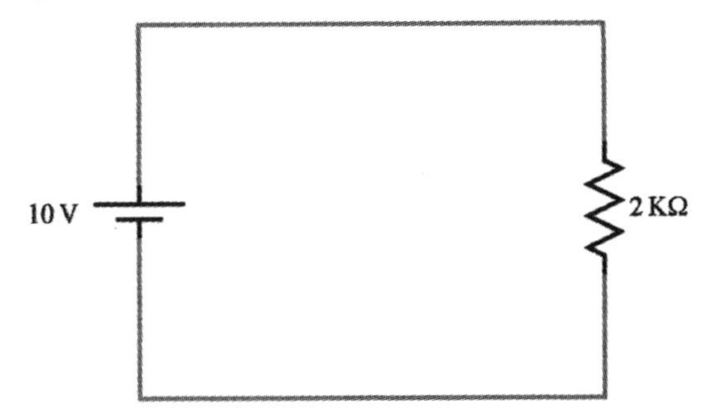

2. Calculate the impedance and the maximum current flow in the following RLC-circuit.

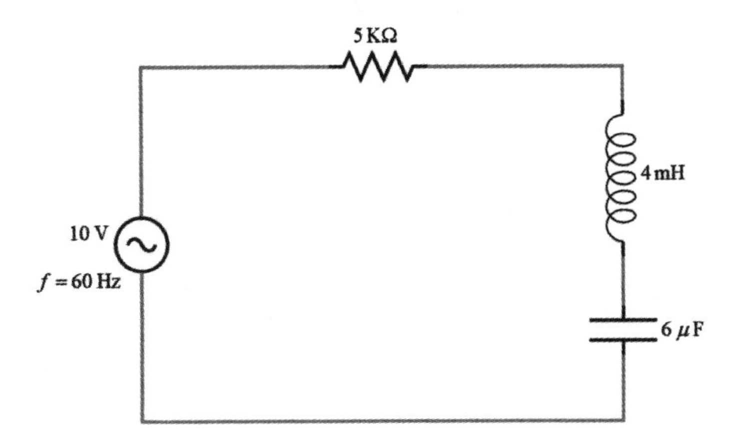

3. What is the resonant frequency of the circuit in Exercise 2?

4. If the power dissipated by the resistor in an RLC-circuit is 5 W, how much energy would it dissipate in the form of heat in 2 seconds?

5. Find the maximum power dissipated by the resistor in the following RLC-circuit:

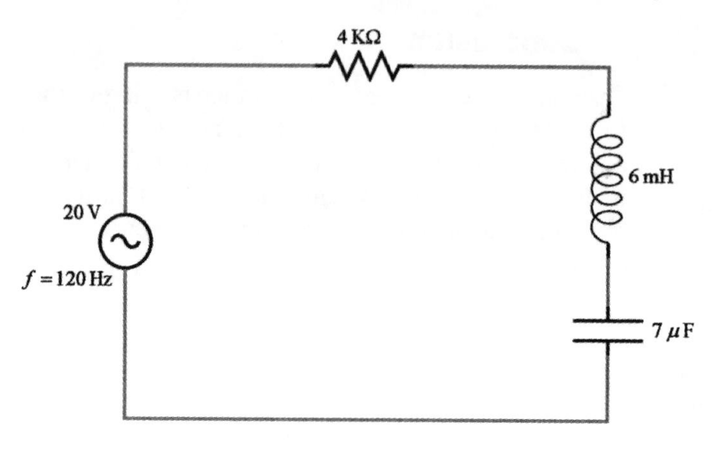

ELECTRONICS 12

The Electromagnetic Signal Produced by a Dipole Antenna

In this application, we apply the wave principles introduced in *College Physics* to the electromagnetic wave produced by a dipole antenna.

In this application, we analyze the origin of the signal produced by a basic dipole antenna and compare it to the transverse waves presented in Chapter 12 of *College Physics*. A physical understanding of the electromagnetic wave produced by an antenna is beneficial for a student of electronics before learning how to modulate the frequency of an electromagnetic wave (FM theory) and/or the amplitude of an electromagnetic wave (AM theory).

A Dipole Antenna

The typical construction of a basic dipole antenna is illustrated in Figure 12.1:

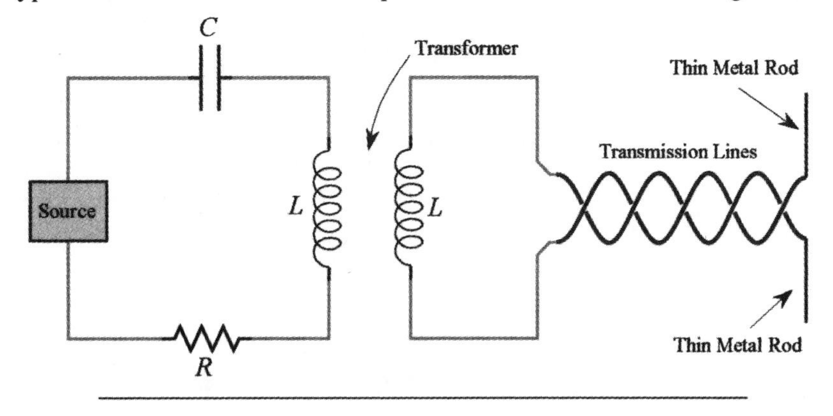

Figure 12.1

The changing magnetic field in the first inductor produces a changing magnetic field in the inductor that is wired to the metal rods. This changing magnetic field in the second inductor produces an alternating current in both the transmission lines and the metal rods. This constantly changing current causes the two antennae to emit an electromagnetic wave with a frequency determined by the frequency of the source voltage, V_{AC}.

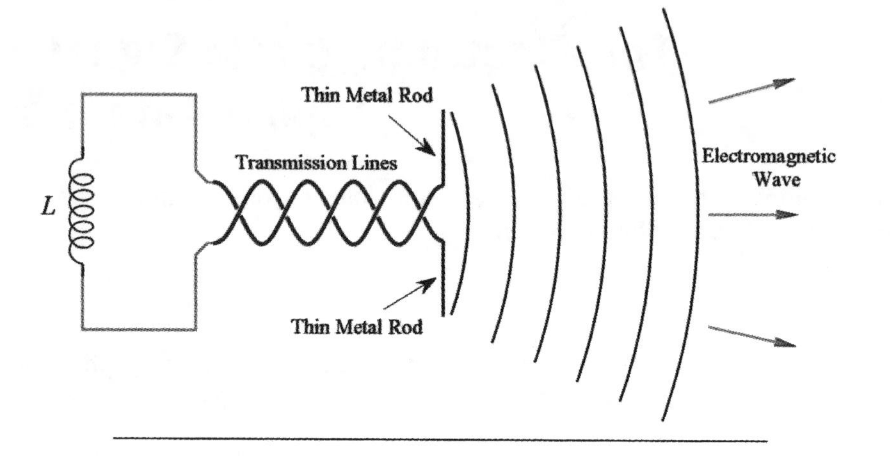

Figure 12.2

This generated electromagnetic wave has two distinct pieces, an electric field that oscillates in the *x*-*y* plane

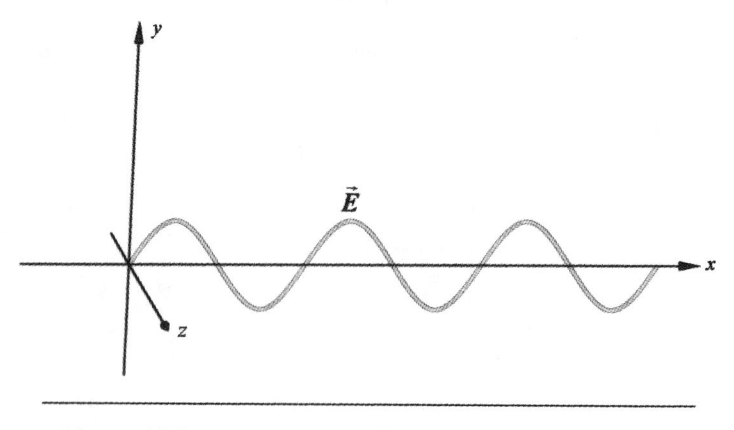

Figure 12.3

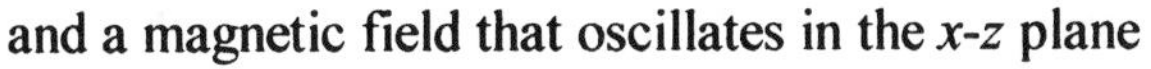
and a magnetic field that oscillates in the x-z plane

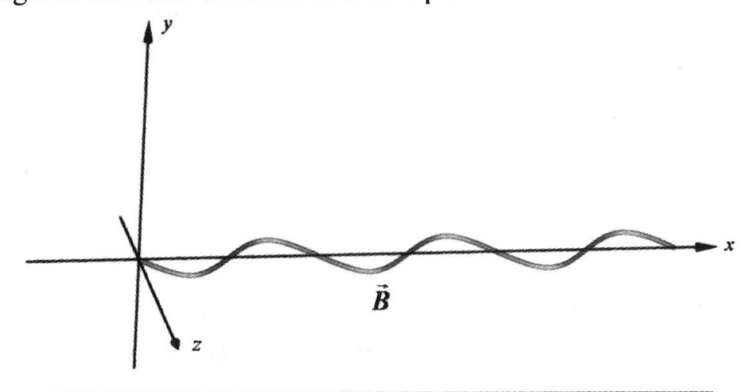

Figure 12.4

> **Connection**
>
> For a discussion of transverse waves, refer to Section 12.3 in *College Physics.*

Notice that while the electric field oscillates along the y-axis and the magnetic field oscillates along the z-axis, the wave propagates along the x-axis. Thus, both the electric and magnetic components are transverse waves. In other words, the electromagnetic wave produced by a dipole antenna is actually two transverse waves that move in phase with one another. They achieve their maximum, minimum, and zero values at the same points.

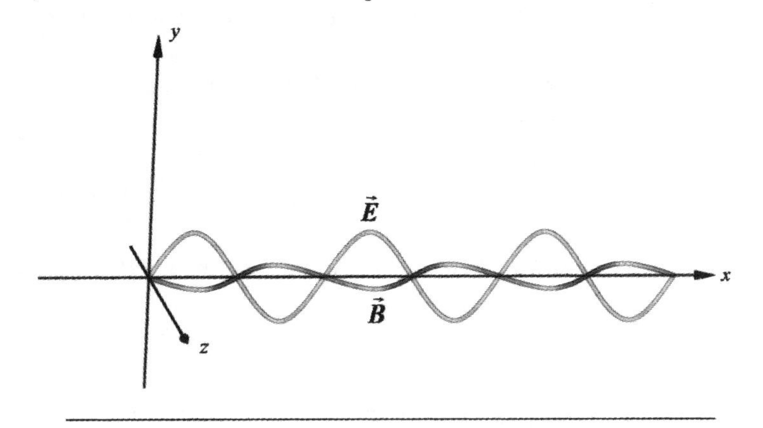

Figure 12.5

These two waves can be expressed using equations that are similar to the equation used to describe a transverse wave on a rope.

$$y = y_m \sin(kx - \omega t) \qquad \textbf{(1)}$$

where

- y is the location of a rope element above or below equilibrium at the time t
- y_m is the amplitude of the wave, the maximum distance that a rope element will be located away from equilibrium due to the wave passing
- k is the angular wave number
- t is the time at which we are analyzing the location of the rope element located at x
- x is the rope element whose location we are finding above or below equilibrium at the time t
- ω is the angular frequency of the wave.

Comparing Equation (1) with the equation that describes the oscillating electric field

$$E = E_m \sin(kx - \omega t)$$

and the equation for the oscillating magnetic field

$$B = B_m \sin(kx - \omega t)$$

reveals that the three equations are remarkably similar. Many of the principles that govern the motion of waves moving along ropes, for example the fact that the velocity of the wave is found by multiplying its wavelength and its frequency, can be extended to the two oscillating fields produced by the antenna.

$y = y_m \sin(kx - \omega t)$	$E = E_m \sin(kx - \omega t)$	$B = B_m \sin(kx - \omega t)$
y is the location of the rope element at the point x at the time t	E is the value of the electric field at the point x at the time t	B is the value of the magnetic field at the point x at the time t
y_m is the maximum distance of a rope element from equilibrium	E_m is the maximum value that the electric field achieves	B_m is the maximum value that the magnetic field achieves
The negative sign in the sine function indicates that the wave is moving to the right	The negative sign in the sine function indicates that the wave is moving to the right	The negative sign in the sine function indicates that the wave is moving to the right
The speed of the wave is found by dividing ω by k, or by multiplying the frequency of the wave by its wavelength	The speed of the wave is found by dividing ω by k, or by multiplying the frequency of the wave by its wavelength	The speed of the wave is found by dividing ω by k, or by multiplying the frequency of the wave by its wavelength

FYI

The speed of light in a vacuum is 2.99792458×10^8 m/s. For most applications, however, it is sufficient to round this value to 3×10^8 m/s.

As an illustration of the connection between the waves on a rope and the electromagnetic waves discussed in this application, let's look at the speed of the waves.

We know that the electric field and the magnetic field are in phase with one another. Consequently, the angular wave number, k, and the angular frequency, ω, must be the same in both the electric field and magnetic field equations. Further, because the electromagnetic waves produced by the antenna propagate at the speed of light, the wave speed equations discussed in Chapter 12 of *College Physics*

$$v = \frac{\omega}{k} \quad \text{and} \quad v = f\lambda$$

can be rewritten as

$$c = \frac{\omega}{k} \quad \text{and} \quad c = f\lambda$$

where the variable c represents the speed of light. Thus, since the speed of light is a constant, once the wave number (angular frequency) is chosen, the angular frequency (wave number) is fixed.

This close relationship between the fields also extends to the field amplitudes. For all electromagnetic waves, regardless of the frequency, the ratio of the amplitudes must always yield the speed of light:

$$c = \frac{E_{\mathrm{m}}}{B_{\mathrm{m}}}$$

Example 1

AM (*Amplitude-Modulated*) radio waves have frequencies that range from 0.53×10^6 Hz to 1.7×10^6 Hz.

Find

a) the range of wavelengths for radio waves
b) the angular frequency of a radio wave possessing a frequency of 0.53×10^6 Hz

Solution

a) If we assume the waves to be moving through a vacuum, we can find the range of associated wavelengths using

$$c = f\lambda$$

For the frequency at the low end of the range, 0.53×10^6 Hz, we calculate

$$c = f\lambda$$

$$\frac{c}{f} = \lambda$$

$$\frac{3\times10^8 \text{ m/s}}{0.53\times10^6 \text{ Hz}} = \lambda$$

$$566 \text{ m} = \lambda$$

Let's pause for a moment and analyze the units in the calculation. Since the unit of hertz is defined as

$$\text{Hz} = \frac{1}{\text{sec}}$$

we can cancel the inverse seconds attached to the frequency with the seconds attached to the speed of light, producing the length unit of meters on the wavelength.

Calculating the frequency at the high end of the range, 1.7 x 10^6 Hz, yields

$$\frac{c}{f} = \lambda$$

$$\frac{3\times 10^8 \text{ m/s}}{1.7\times 10^6 \text{Hz}} = \lambda$$

$$176.5 \text{ m} = \lambda$$

The relationship between linear and angular frequency can be found in Section 12.2 of *College Physics.*

b) The angular frequency associated with the linear frequency of 0.53 x 10^6 Hz can be found using

$$\omega = 2\pi f$$

$$\omega = 2\pi\left(0.53\times 10^6 \frac{1}{\text{s}}\right)$$

$$\omega = 3.33\times 10^6 \frac{\text{rad}}{\text{s}}$$

Example 2

An antenna produces an electromagnetic wave that travels through a vacuum. If the electric field of the wave has an amplitude of 1000 N/C, find the amplitude of the associated magnetic field.

Solution Using the relationship between the field amplitudes, we insert the values for the speed of the wave and the electric field amplitude:

Connection

The units of velocity, electric fields, and magnetic fields can be found in the inside front cover of this text, or *College Physics.*

$$c = \frac{E_{\text{m}}}{B_{\text{m}}}$$

$$B_{\text{m}} = \frac{E_{\text{m}}}{c}$$

$$B_{\text{m}} = \frac{1000 \text{ N/C}}{3\times 10^8 \text{ m/s}}$$

$$B_{\text{m}} = 3.33\times 10^{-6} \text{ T}$$

Notice that in this example we used the SI standard units of newtons per coulomb for the electric field amplitude and tesla for the magnetic field amplitude. Although we have opted for this set of units in which to execute our calculation, it is important to note that the relationship between the electric field amplitude and the magnetic field amplitude is independent of the choice of units.

Conclusion

From televisions to cellular phones, electromagnetic signals from satellites and other sources play an integral part in our lives. Because of the many and varied applications in electronics and electronic communications of these waves, an understanding of the behavior of electromagnetic signals is important for any student pursuing an electronics career.

Exercises

1. For the wave given by

$$y = 10\sin(4x - 2t)$$

find

   a) the amplitude of the wave
   b) the angular frequency of the wave
   c) the angular wave number
   d) the linear frequency of the wave
   e) the wavelength
   f) the speed of the wave
   g) the direction of the wave's travel

2. An electromagnetic wave moving in a vacuum has an angular frequency of 1.2×10^{15} rad/sec. Find the angular wave number of this wave.

3. The electromagnetic wave corresponding to the color yellow-green has a wavelength of 550 nm.

   a) What is the linear frequency of this wave?
   b) What is the angular wave number of this wave?

4. In the following antenna diagram, the source circuit is operating at resonant frequency. What is the frequency of the alternating current in the transmission lines?

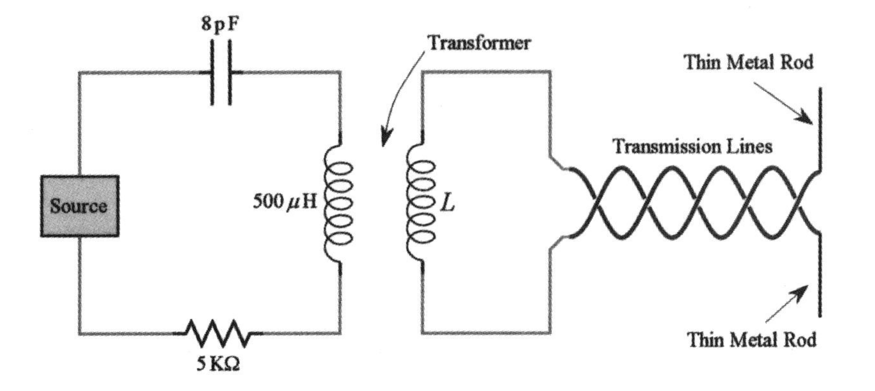

5. If the magnitude of the oscillating electric field for a wave is 2000 N/C, what is the magnitude of the oscillating magnetic field for this wave?

6. In FM (*Frequency Modulation*) theory, the information signal is superposed on top of the constant-frequency carrier signal. This superposition modifies the constant frequency of the carrier wave. On the receiving end, the frequency of the carrier wave is removed from the transmitted signal, thus leaving the electromagnetic wave that corresponds to the information being transferred. Discuss the response of the angular wave number of the wave due to this frequency modulation.

7. A driver enters a tunnel under a large body of water. As she searches for radio stations, she finds that she is still able to receive some stations, but others yield only static.

   a) Carry out the necessary research to determine which types of waves, AM or FM, are more likely to be received in a tunnel. Does it matter whether the tunnel is below water or solid earth? Why?
   b) Explain why certain types of radio waves can be received while others yield only static.

ELECTRONICS 13

Diffraction and Television Signals

In this short application, we use the phenomenon of diffraction to explain how television and radio signals can be received in valleys.

> **Connection**
>
> Diffraction is introduced in Section 14.3 of *College Physics.*
>
> Huygen's Principle is discussed in Section 13.3.

The principle of diffraction, used in conjunction with the concepts expressed by Huygen's Principle, is useful in helping us to understand how a house located in a valley can still receive television and radio signals from a source situated outside of the valley.

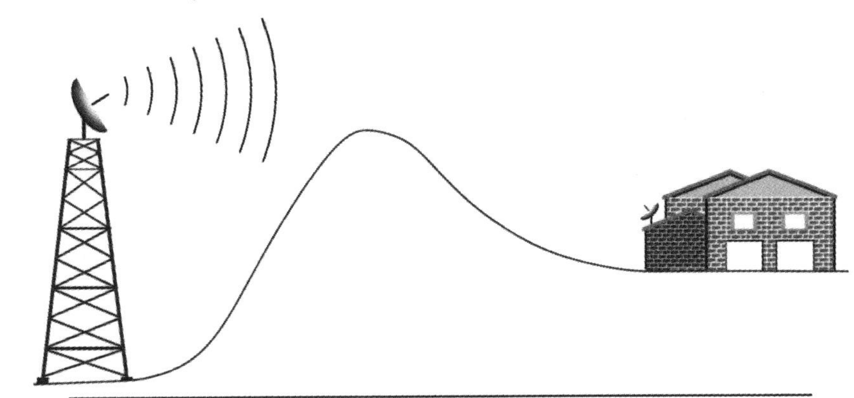

Figure 13.1

> 
>
> **Connection**
>
> See Electronics Application 12 - *The Electromagnetic Signal Produced by a Dipole Antenna* for a discussion of the creation of an electromagnetic wave by an antenna.

Diffraction

The electromagnetic wave generated by the antenna shown in Figure 13.1 spreads as it moves away from the antenna.

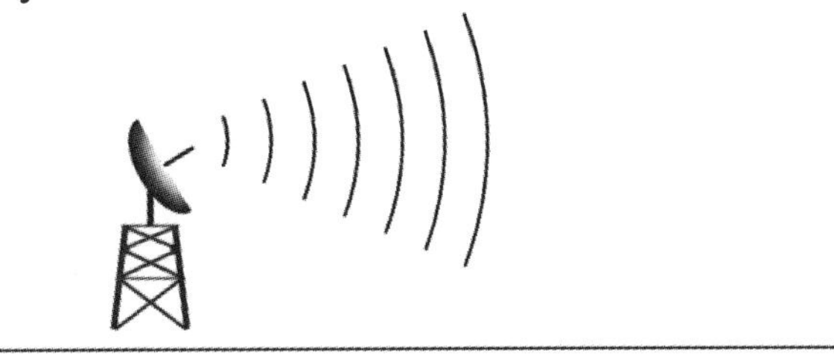

Figure 13.2

When the wave hits the hillside, the top of the hill functions as a diffraction grating and becomes the source of a new electromagnetic wave.

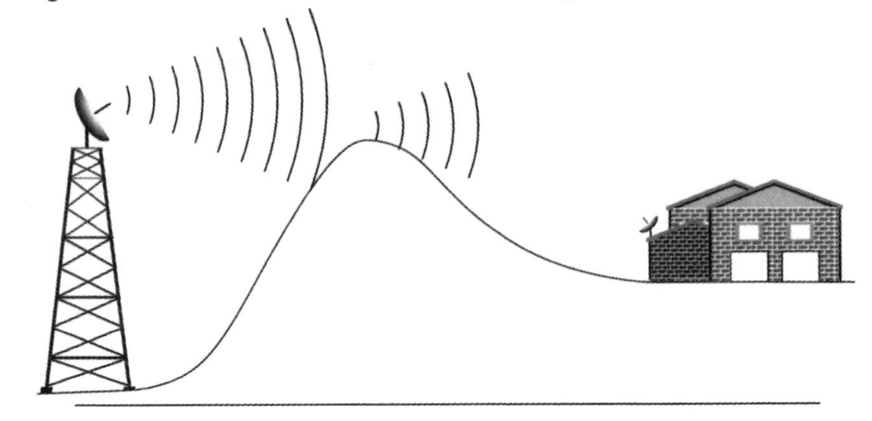

Figure 13.3

This new wave, in accordance with Huygen's Principle, radiates outward as it moves away from the hill and thus intersects the receiving antenna attached to the house.

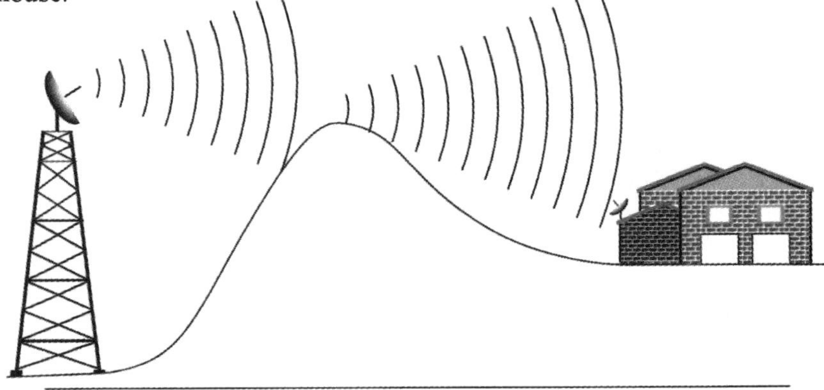

Figure 13.4

Because the wave must move a certain distance before it has spread out enough to touch ground, a region close to the hill will not receive any wavefronts. This region is called the *shadow zone* and is discussed in Section 14.3 of *College Physics*.

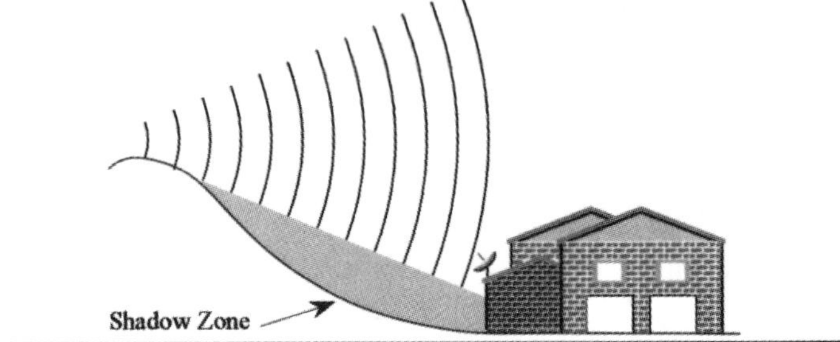

Figure 13.5

▸ Conclusion

An understanding of diffraction is necessary for a full understanding of electronic communications. A student wishing to pursue this field of study must understand the nature and behavior of the electromagnetic signals produced by the antennae and satellites that will be part of his or her career. Because of the mountains, skyscrapers, etc., in our world, a technician analyzing these electromagnetic signals must have a grasp of the physical phenomenon of diffraction in order to predict how these signals are going to behave.

Exercises

1. What might be some ways that we can decrease the size of the shadow zone in the valley, thus increasing the number of homes that receive the signal?

2. Use the material from this application, as well as the material on diffraction in *College Physics*, to explain how it is possible to hear but not see around corners.

ELECTRONICS 14

Relativistic Kinetic Energy and Momentum

In this application, we use the motion of a conduction electron in an electrical circuit to illustrate the limit of applicability of the Newtonian Laws of Motion.

Connection

Kinetic energy is introduced in Section 5.3 of *College Physics*.

Linear momentum is introduced in Section 6.1 of *College Physics*.

In 1905, Albert Einstein astounded the scientific world by proposing a completely new way of analyzing the motion of objects. His new theory, Special Relativity, put forth that not only is the measured velocity of an object dependent on the relative velocity between the reference frames of the observer and the object, but also that there is a universal upper bound on the velocity that an object can possess. This upper bound is the velocity of a light wave in a vacuum.

The rules of Special Relativity, now generally accepted and used in laboratories daily, are used to describe accurately the motion of objects that are moving at high speeds. It is now known that the laws of Newtonian mechanics are low-speed approximations of their more accurate counterparts from Special Relativity.

Thus, when analyzing such physical characteristics as the kinetic energy and the momentum of high-speed electrons in a circuit, the more accurate rules of Special Relativity should be applied.

The Lorentz Factor

As we mentioned, Special Relativity incorporates the possible velocity that may exist between the reference frame of the observer and the reference frame of the object being analyzed. This relative velocity, v, can be used to calculate a numerical coefficient that is incorporated into many of the calculations in Special Relativity. This coefficient is called the *Lorentz factor* (γ), and is given by the equation

$$\gamma = \frac{1}{\sqrt{1 - \frac{v^2}{c^2}}}$$

where v is the relative velocity between the reference frames of the observer and the system being measured, and c is the speed of light in a vacuum:

$$c = 2.99792458 \times 10^8 \text{ m/s}$$

Notice that because both v and c are velocities, the units in the denominator of the Lorentz factor will cancel. Thus, the Lorentz factor is a dimensionless, numerical coefficient that can be used to modify the equations for several of the Newtonian concepts discussed in *College Physics*, such as momentum and kinetic energy. With the introduction of the Lorentz factor, these equations will then correctly describe objects that are moving with large velocities.

Example 1

Calculate the Lorentz factor for an electron moving past an observer at 80% of the speed of light.

Solution Although it is actually possible to calculate 80% of the speed of light as the quantity to be inserted for the velocity in the equation for the Lorentz factor, it is preferable simply to express the velocity as

$$v = 0.8c$$

The calculation of the Lorentz factor then becomes

$$\gamma = \frac{1}{\sqrt{1 - \frac{v^2}{c^2}}}$$

$$\gamma = \frac{1}{\sqrt{1 - \frac{(0.8c)^2}{c^2}}}$$

$$\gamma = \frac{1}{\sqrt{1 - \frac{0.64c^2}{c^2}}}$$

$$\gamma = \frac{1}{\sqrt{1 - 0.64}}$$

$$\gamma = \frac{1}{\sqrt{0.36}}$$

$$\gamma = \frac{1}{0.6}$$

$$\gamma = 1.67$$

Relativistic Momentum

Connection

Linear momentum is discussed in Section 6.1 of *College Physics.*

We know that the Newtonian expression for the momentum of a mass is given by

$$\vec{p} = m\vec{v}$$

However, all of the measurements of momenta that are carried out in Chapter 6 of *College Physics* assume that we are external observers as the various masses move past us. In other words, we are not in the same reference frame as the moving objects that we are analyzing. It is not surprising then that the Newtonian expression given above will need to be modified to incorporate this discrepancy in reference frames.

We can make the modification easily by inserting the Lorentz factor that incorporates the relative velocity between the reference frame of the observer and the reference frame of the moving mass.

Newtonian Momentum	$p = mv$
Relativistic Momentum	$p = \gamma mv$

If we write out the Lorentz factor in detail, the relativistic momentum becomes

$$p = \left(\frac{1}{\sqrt{1 - \frac{v^2}{c^2}}} \right) mv$$

Notice that as the velocity between the reference frames becomes larger, the denominator in the Lorentz factor becomes *smaller*. This means that this numerical coefficient will become larger as the relative velocity between the reference frames grows. It follows then that the larger the velocity of the object, the larger the difference between the Newtonian expression and the relativistic one. Therefore, for high-speed objects, such as electrons in circuits, a very large error can result if the Newtonian expression is used instead of its more accurate relativistic counterpart.

Example 2

Calculate the relativistic momentum of an electron moving past an observer at 80% of the speed of light.

Solution From Example 1, we know that the Lorentz factor between the reference frame of the observer and the reference frame of the electron is

$$\gamma = 1.67$$

Inserting this value along with the mass and velocity of the electron yields

$$p = \gamma mv$$
$$p = (1.67)(9.11 \times 10^{-31}\ \text{Kg})(0.8)(2.99792458 \times 10^{8}\ \text{m/s})$$
$$p = 3.65 \times 10^{-22}\ \text{Kgm/s}$$

In comparison, the Newtonian calculation yields the different result of

$$p = mv$$
$$p = (9.11 \times 10^{-31}\ \text{Kg})(0.8)(2.99792458 \times 10^{8}\ \text{m/s})$$
$$p = 2.18 \times 10^{-22}\ \text{Kgm/s}$$

Thus, the difference between the Newtonian concept of momentum and the relativistic one depends upon the velocity of the object with respect to the person measuring the momentum. For large velocities, such as the velocity of electrons in circuits, the difference is sufficiently significant to require that we take it into consideration when analyzing the individual electrons that comprise an electric current.

Relativistic Kinetic Energy

Because Einstein's Theory of Special Relativity also radically altered our understanding of mass, energy, and the interplay between them, the move from the Newtonian view of kinetic energy to the relativistic one is much more extensive than is its momentum counterpart.

In the low-speed Newtonian view, mass is the *amount* of material and energy is a *property* of that material. In the relativistic view, this distinction is blurred. Mass and energy are viewed simply as two different facets of the *one* physical quantity, *mass-energy*.

Where Newtonian kinetic energy is given by

$$\text{KE} = \frac{1}{2}mv$$

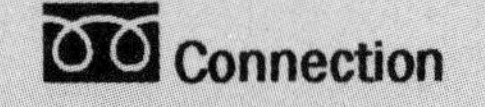

Kinetic Energy is discussed in Section 5.3 of *College Physics*.

relativistic kinetic energy is given by the much more elaborate equation

$$\mathrm{KE} = mc^2(\gamma - 1)$$

Although it might not be apparent at first glance, it is actually possible to begin with the more accurate relativistic kinetic energy equation and recover its low-speed counterpart using the appropriate mathematics. However, this level of mathematics is outside of the scope of this application since our purpose here is to discuss the limitations of the Newtonian theory and to illustrate how Special Relativity relates to electronics.

Example 3

Calculate the relativistic kinetic energy of an electron moving past an observer at 80% of the speed of light.

Solution Again using the Lorentz factor calculated in Example 1, the relativistic kinetic energy becomes

$$\mathrm{KE} = mc^2(\gamma - 1)$$

$$\mathrm{KE} = (9.11\times10^{-31}\ \mathrm{Kg})(2.99792458\times10^{8}\ \mathrm{m/s})^2(1.67 - 1)$$

$$\mathrm{KE} = 5.48\times10^{-14}\ \mathrm{J}$$

Notice that the relativistic expression has produced the same energy unit of joules as the non-relativistic case discussed in *College Physics*:

$$\frac{\mathrm{Kgm}^2}{\mathrm{s}^2} = \frac{\mathrm{Kgm}}{\mathrm{s}^2}\cdot\mathrm{m} = \mathrm{N}\cdot\mathrm{m} = \mathrm{J}$$

Conclusion

In summary, although the differences between the Newtonian methods of analyzing the motion of objects and the relativistic methods are sometimes small, it is important to realize that the Newtonian methods have a limit on their validity. The closer the velocity of an object gets to the universal upper bound of the speed of light, the more important it becomes for us to use the more accurate methods of Special Relativity. Therefore, since all electronics is based upon the motion of electrons, there may be times when an electronics technician or an electrical engineer may be forced to perform his or her calculations using the more accurate techniques of Special Relativity instead of the low-speed approximations of Newtonian physics.

For Further Thought

A familiar equation can be found by coupling the expression for the relativistic kinetic energy of a mass

$$\text{KE} = mc^2(\gamma - 1)$$

with the relativistic expression for the *total* energy of a mass

$$E = \gamma mc^2$$

Beginning with the kinetic energy expression, we see

$$\begin{aligned} \text{KE} &= mc^2(\gamma - 1) \\ \text{KE} &= mc^2\gamma - mc^2 \\ \text{KE} &= \gamma mc^2 - mc^2 \\ \text{KE} + mc^2 &= \gamma mc^2 \end{aligned}$$

Because the right-hand side of our result is the same as the right-hand side of the total energy expression, we can write the total energy as

$$E = \text{KE} + mc^2$$

Review

Remember that a mass must be moving to have kinetic energy.

In the case of a stationary mass, the kinetic energy becomes zero. Thus, the total relativistic energy becomes

$$\begin{aligned} E &= \text{KE} + mc^2 \\ E &= 0 + mc^2 \\ E &= mc^2 \end{aligned}$$

The result is Einstein's famous equation that relates energy and mass. This equation states that mass and energy are equivalent and are simply different forms of one another. It is this deeper understanding of the relationship between mass and energy, produced by Special Relativity, that gave birth to the present nuclear power industry.

Exercises

1. Calculate the Lorentz factor for a particle moving at 50% of the speed of light.

2. Calculate the relativistic kinetic energy and momentum of an electron moving at 85% of the speed of light.

3. Calculate the relativistic kinetic energy and momentum of a proton moving at 90% of the speed of light. The mass of a proton is 1.67×10^{-27} Kg.

4. Given that the Lorentz factor of an electron is 1.5, calculate the electron's relativistic kinetic energy and momentum.

5. Discuss the effect on an electron's kinetic energy if we could actually accelerate it to the speed of light.

ELECTRONICS 15

Physics and the Next Generation of Computing

In this application, we discuss how the laws of physics are generating an exciting new area of research in the computer industry.

The Computing Challenge

Researchers in the computer industry face the constant challenge of how to make faster and more powerful computers to handle the massive amount of information produced and the expectation for instant communication and analysis. From the early modern-day computers that used relays and transistors to today's silicon chips with micron-wide logic gates, developers have worked for the last 50 years to decrease the size and increase the speed of the processors required for data storage, manipulation, and communication. This race to build better and faster computers has created a new problem for researchers who must confront objects at the size scale of individual atoms, a scale at which Newtonian physics no longer applies.

Connection

It may be helpful to review Electronics Application 14 - *Relativistic Kinetic Energy and Momentum.*

In Application 14, *Relativistic Kinetic Energy and Momentum*, we saw that the Newtonian formulation of physics has a limited range of applicability. In that application, we discussed the fact that for high-speed objects, the Newtonian formulation is inadequate. In this application, we will continue this line of thought to discuss the limitations of the Newtonian formulation in the world of the very small. The breakdown of the Newtonian formulation in the sub-atomic world led researchers to a deeper understanding of our world and eventually helped them to produce most of the electronic devices currently used in technology and industry. In this application, our focus is on the fascinating research being conducted to use knowledge of the sub-atomic world to produce the next generation of computers.

The Quantum Problem

At the end of the 19th Century, experiments were already generating results that were inconsistent with the Newtonian predictions. These experiments, conducted at the size scale of atoms and electrons, yielded results that were radically different from those predicted by Newtonian mechanics. For the first time since its formulation, the Newtonian theory of physics was not successful in predicting the location and velocity of the objects to which it was being applied. Because a physical theory must agree with and produce the same results as a physical experiment to be deemed true, the Newtonian theory came into question as the "theory of all objects."

> **Connection**
>
> Energy is discussed in Chapter 5 of *College Physics.*

For example, researchers soon observed that the particles being analyzed, such as electrons, did not execute the simple, smooth trajectories familiar from classical physics. Work done by Albert Einstein, Max Planck, and others, led scientists to believe that the energies possessed by the electrons only came in multiples of a certain fundamental amount, or *quantum.* An electron could have, for example, 5 times an amount of energy or 6 times the amount, but it could not have 5.5 times the amount. It could have 5 times the base amount, or 6, but not a value *in between* these integer multiples. This realization, in contrast to the tenets of classical physics, led to the understanding that there is not a continuum of values for the possible amount of energy possessed by the electron. In the language of this new theory, the energy is said to be *quantized.*

Although we know it to exist, this *quantization* is not apparent in the macroscopic world. A useful analogy is to think about how we see a movie as being continuous, even though it is not. A movie actually consists of many individual frames that appear as a smooth image because of the speed at which the frames move.

Quantization of Electronic Orbitals

In a related development, researchers found that energy was not the only property of electrons that appeared to come in discrete amounts, or *quanta.* Another example of *quantization* arose during an effort to explain electronic orbitals. The Danish physicist Niels Bohr put forth a theory incorporating the quantum concept to explain why electrons do not spiral into the atomic nucleus.

According to classical physics, when an electric charge accelerates, energy is released in the form of an electromagnetic wave. Because the electrons appear to execute circular orbits, they should experience a centripetal force and therefore a centripetal acceleration. Since the electrons are constantly being accelerated centripetally, they should release electromagnetic radiation that will decrease their energy and result in their spiraling into the nucleus of the atom. However, we know this scenario is untrue because we know stable atoms exist. Using the concept of quantization, Bohr was able to provide an explanation for this apparent paradox.

Connection

Centripetal force is discussed in Section 7.3 of *College Physics*.

In a radical step, Bohr postulated that electrons cannot exist at all locations surrounding the nucleus. Electrons are only allowed to exist in discrete locations called *orbitals*. An electron can exist in one orbital or another, *but not in between orbitals*. Thus, as in the case of the electron's energy, electronic orbitals are also quantized.

The Creation of Quantum Mechanics

Based on such results as the quantization of energies and the quantization of electronic orbitals, researchers soon realized that a simple modification of the Newtonian formulation was not going to be sufficient. Because all of the physical properties of the objects being studied appeared to be quantized instead of continuous, a radical new theory was required to describe correctly the world of the very small. This new theory would eventually be called *quantum mechanics*.

However, the theory of quantum mechanics developed by Werner Heisenberg, Pascual Jordan, and Erwin Schrodinger during the 1920s, that correctly described the quantum behavior of the sub-atomic world, also made physical predictions that were at once startling and philosophically baffling. One of the predictions is the *principle of superposition.*

Superposition

According to the theory of quantum mechanics, not only are the allowed physical states of a particle quantized, the particle actually exists in several of these states *at the same time*. Amazingly, the theory says that it is possible for a particle such as an electron to exist in a *superposition* of the various possible values of a physically measurable quantity such as energy or position. The act of a person measuring the system causes the electron to "collapse" into one of the possible states.

When first confronted with this concept, a common reaction is that the electron, or other particle, was always in a precise state but that we just had not looked at it yet. In actuality, this simultaneous existence in several states appears to be true and calculations that ignore this superposition do not agree with experimental results. Interestingly, this strange quantum phenomenon may provide a solution to the problem of creating the next generation of faster computers.

Superposition and Parallel Computing?

In today's digital computers, information is coded using a combination of 1's and 0's. The smallest piece of information, called a *bit*, is specified using either a zero or a one. Thus, a word or other piece of information that contains *n* bits, is expressed using a string of *n* zeros and ones. For example, in a typical digital register, three bits are required to express each of the numbers 1 through 8; 000, 001, 010, etc.

FYI

The *spin* of an electron is related to its angular momentum.

Angular momentum is discussed in Section 8.5 of *College Physics*.

A new area of research has developed to improve upon the currently used bit concept. Computer researchers are currently attempting to develop a radical new computing entity - the quantum bit, or *qubit*. These qubits may be represented by an atom or particle that possesses one of two states. For example, when measuring the *spin* of an electron, it only has one of two possible values, up or down. In other words, the spin of the electron is quantized like the other physical properties in the microscopic world. These two spin states can be assigned values of 1 and 0 and the electron can now be used as a qubit. Thus, two electrons can describe four possible states, and three electrons can describe eight possible states, just as does a classical digital register.

However, unlike the classical register that can only represent one of the numbers between zero and eight at a time, a system of qubits has access to each of the numbers *simultaneously*. Because of the principle of superposition, quantum bits do not exist in a single state, but rather in a superposition of all their possible states. Therefore, in some sense a register consisting of qubits exists in all of the possible states between zero and eight simultaneously. This simultaneity may have a profound impact on parallel computing since a computer that used qubits would have information about all of the possible states at the same time.

Quantum Entanglement and "Wiring" a Computer that Uses Qubits

In addition to providing access to huge amounts of information simultaneously, a processor that used qubits may have a built-in method of "wiring" them together. This built-in system involves another unusual aspect of the microscopic world known as *quantum entanglement*.

Connection

The electric field of a light wave is discussed in Electronics Application 12 - *The Electromagnetic Signal Produced by a Dipole Antenna.*

A visual example of quantum entanglement can be constructed using two quanta of light, or *photons*. Suppose that we construct an experiment in which we use two photons (particles of light) that have opposite polarizations for their electric fields. We then shoot one photon to the left and one photon to the right. Until an observer measures the polarization of one of the photons, the two polarizations remain unknown. However, once the polarization of one of the photons is measured, the polarization of the other photon is immediately known since we know that the other photon must have an opposite polarization to the one we measure. Amazingly, this relationship between the two photons exists *no matter how large the distance between the photons*! Regardless of the distance between the photons or the length of time they are separated, a physical connection exists between them.

This type of quantum entanglement could also be applied to qubits. If the state of two or more qubits were known, they would always maintain a connection with the other qubits with which they were "entangled" at the start of the process even as they move and carry out their designed behavior.

Conclusion

Although the concept of quantum computing may be far from the point of desktop use, it does provide an example of how physics is at play in all areas of technology. An understanding of physics provides today's technician with a powerful tool for career development. In this application, we focused on how the laws of physics have placed the computer industry at the edge of a very exciting revolution in technology. However, almost all areas of electronics and electronic engineering are constantly being upgraded and developed using the laws of both classical and quantum physics.

Exercises

1. Use the concept of *hysteresis*, familiar from electronics, to explain how information is stored on magnetic tapes.

2. Execute the necessary research, and explain how computers are able to take the sentences entered on a keyboard and translate them into digital information.

3. Execute the necessary research to explain how files that contain large amounts of information are "zipped."

Mathematical Symbols

$=$	is equal to
$\neq$	is not equal to
$\approx$	is approximately equal to
$\sim$	about
$\propto$	is proportional to
$>$	is greater than
$\geq$	is greater than or equal to
$\gg$	is much greater than
$<$	is less than
$\leq$	is less than or equal to
$\ll$	is much less than
$\pm$	plus or minus
$\mp$	minus or plus
$\bar{x}$	average value of x
Δx	change in x
$\|x\|$	absolute value of x
Σ	sum of
∞	infinity

The Greek Alphabet

Alpha	Α	α	Nu	Ν	ν
Beta	Β	β	Xi	Ξ	ξ
Gamma	Γ	γ	Omicron	Ο	o
Delta	Δ	δ	Pi	Π	π
Epsilon	Ε	ε	Rho	Ρ	ρ
Zeta	Ζ	ζ	Sigma	Σ	σ
Eta	Η	η	Tau	Τ	τ
Theta	Θ	θ	Upsilon	Υ	υ
Iota	Ι	ι	Phi	Φ	ϕ
Kappa	Κ	κ	Chi	Χ	χ
Lambda	Λ	λ	Psi	Ψ	ψ
Mu	Μ	μ	Omega	Ω	ω

Quadratic Formula

If $ax^2 + bx + c = 0$, then

$$x = \frac{-b \pm \sqrt{b^2 - 4ac}}{2a}$$

Values of Some Useful Numbers

$\pi = 3.141\,59\ldots$ $\sqrt{2} = 1.414\,21$

$e = 2.718\,28\ldots$ $\sqrt{3} = 1.732\,05$

Trigonometric Relationships

Definitions of Trigonometric Functions

$$\sin\theta = \frac{y}{r} \qquad \cos\theta = \frac{x}{r} \qquad \tan\theta = \frac{\sin\theta}{\cos\theta} = \frac{y}{x}$$

$\theta°$ (rad)	$\sin\theta$	$\cos\theta$	$\tan\theta$
0° (0)	0	1	0
30° ($\pi/6$)	0.500	$\sqrt{3}/2 \approx 0.866$	$\sqrt{3}/3 \approx 0.577$
45° ($\pi/4$)	$\sqrt{2}/2 \approx 0.707$	$\sqrt{2}/2 \approx 0.707$	1.00
60° ($\pi/3$)	$\sqrt{3}/2 \approx 0.866$	0.500	$\sqrt{3} \approx 1.73$
90° ($\pi/2$)	1	0	∞

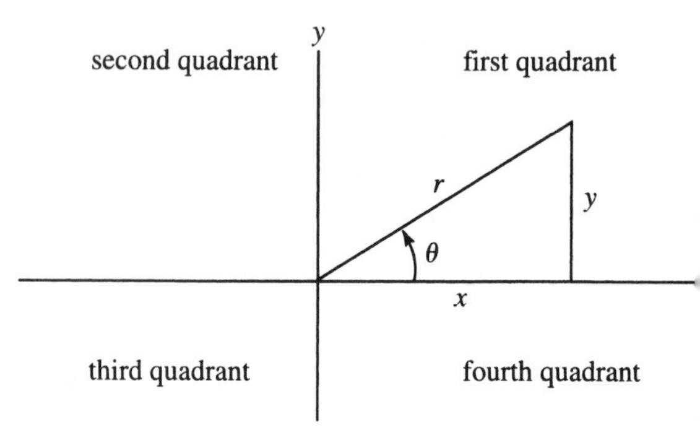